Advances in Intelligent Systems and Computing

Volume 855

Series editor

Janusz Kacprzyk, Systems Research Institute, Polish Academy of Sciences, Warsaw, Poland
e-mail: kacprzyk@ibspan.waw.pl

The series "Advances in Intelligent Systems and Computing" contains publications on theory, applications, and design methods of Intelligent Systems and Intelligent Computing. Virtually all disciplines such as engineering, natural sciences, computer and information science, ICT, economics, business, e-commerce, environment, healthcare, life science are covered. The list of topics spans all the areas of modern intelligent systems and computing such as: computational intelligence, soft computing including neural networks, fuzzy systems, evolutionary computing and the fusion of these paradigms, social intelligence, ambient intelligence, computational neuroscience, artificial life, virtual worlds and society, cognitive science and systems, Perception and Vision, DNA and immune based systems, self-organizing and adaptive systems, e-Learning and teaching, human-centered and human-centric computing, recommender systems, intelligent control, robotics and mechatronics including human-machine teaming, knowledge-based paradigms, learning paradigms, machine ethics, intelligent data analysis, knowledge management, intelligent agents, intelligent decision making and support, intelligent network security, trust management, interactive entertainment, Web intelligence and multimedia.

The publications within "Advances in Intelligent Systems and Computing" are primarily proceedings of important conferences, symposia and congresses. They cover significant recent developments in the field, both of a foundational and applicable character. An important characteristic feature of the series is the short publication time and world-wide distribution. This permits a rapid and broad dissemination of research results.

More information about this series at http://www.springer.com/series/11156

Raquel Fuentetaja Pizán
Ángel García Olaya · Maria Paz Sesmero Lorente
Jose Antonio Iglesias Martínez
Agapito Ledezma Espino
Editors

Advances in Physical Agents

Proceedings of the 19th International
Workshop of Physical Agents (WAF 2018),
November 22–23, 2018,
Madrid, Spain

 Springer

Editors
Raquel Fuentetaja Pizán
Computer Science Department
Universidad Carlos III de Madrid
Leganés, Madrid, Spain

Jose Antonio Iglesias Martínez
Computer Science Department
Universidad Carlos III de Madrid
Leganés, Madrid, Spain

Ángel García Olaya
Computer Science Department
Universidad Carlos III de Madrid
Leganés, Madrid, Spain

Agapito Ledezma Espino
Computer Science Department
Universidad Carlos III de Madrid
Leganés, Madrid, Spain

Maria Paz Sesmero Lorente
Computer Science Department
Universidad Carlos III de Madrid
Leganés, Madrid, Spain

ISSN 2194-5357 ISSN 2194-5365 (electronic)
Advances in Intelligent Systems and Computing
ISBN 978-3-319-99884-8 ISBN 978-3-319-99885-5 (eBook)
https://doi.org/10.1007/978-3-319-99885-5

Library of Congress Control Number: 2018961590

This Springer imprint is published by the registered company Springer Nature Switzerland AG
The registered company address is: Gewerbestrasse 11, 6330 Cham, Switzerland

Preface

Physical agents are likely to change the society in the next years. Despite their inherent risks, many times exaggerated by apocalyptic and catastrophic views, it is expected that they will revolutionize our daily life and will have a positive impact in many different fields, from work to health or leisure. The possibilities for research and application are limitless, and the unstoppable "robotics revolution" will increase well-being and open a myriad of new opportunities and challenges.

The International Conference Workshop of Physical Agents (WAF 2018) is a forum for information and experiences exchange in different areas regarding the concept of *agent* on physical environments, especially applied to the control and coordination of autonomous systems: robots, mobile robots, industrial processes, or complex systems. The authors cover several topics in different areas such as software agents, multiagent systems, robotic manipulators, RoboCup and soccer robots, autonomous and semiautonomous robots, machine learning and robotics, industrial robotics, computer vision and robotics, artificial vision and robotics, and artificial intelligence and robotics.

The nineteenth edition of the conference has been organized at Madrid in November 2018 by the Department of Computer Science of the Universidad Carlos III de Madrid, Spain, and the Spanish Physical Agents Network (*Red de Agentes Físicos*), and technically sponsored by Robotics journal and Springer.

This volume collects 22 papers (73 authors) accepted and presented at the conference. The 73 authors from 5 different countries confirm the international status of the event.

This conference will provide a friendly atmosphere and will be a leading international forum focusing on discussing problems, research, results, and future directions in the area of *physical agents*.

WAF 2018 has been possible thanks to the work of many people. We would like to thank the authors and reviewers. Thanks to the Universidad Carlos III de Madrid for letting us use their facilities for the conference sessions. Thanks for the hard work and dedication of the program and organizing committee members. And special thanks to our editor, Springer, that is in charge of this Conference Proceedings edition. Thank you.

September 2018   Angel Garcia-Olaya
Raquel Fuentetaja
Jose Antonio Iglesias
Agapito Ledezma
M. Paz Sesmero Lorente

Organization

Program Committee

Eugenio Aguirre	University of Granada, Spain
José María Armingol	Universidad Carlos III de Madrid, Spain
Ivan Armuelles Voinov	University of Panama, Panamá
George Azzopardi	University of Groningen, Netherlands
Pilar Bachiller	Universidad de Extremadura, Spain
Antonio Bandera	University of Malaga, Spain
Juan Pedro Bandera Rubio	University of Malaga, Spain
Rafael Barea	University of Alcalá, Spain
Luis M. Bergasa	University of Alcalá, Spain
Alberto Bugarín	University of Santiago de Compostela, Spain
Pablo Bustos	Universidad de Extremadura, Spain
Luis V. Calderita Estévez	Universidad de Extremadura, Spain
Oscar Camacho	Escuela Politécnica Nacional, Ecuador
Miguel Cazorla	Universidad de Alicante, Spain
José María Cañas Plaza	Universidad Rey Juan Carlos, Spain
Juan Carlos Corrales	Universidad del Cauca, Colombia
Fernando Fernandez	Universidad Carlos III de Madrid, Spain
Antonio Fernández-Caballero	Universidad de Castilla-La Mancha, Spain
Miguel Angel Garcia	Universidad Autónoma de Madrid, Spain
Miguel Garcia-Silvente	University of Granada, Spain
Ismael Garcia-Varea	Universidad de Castilla-La Mancha, Spain
Antonio González	University of Granada, Spain
Victor González-Castro	Universidad de León, Spain
Jose Daniel Hernández Sosa	Universidad de las Palmas de Gran Canaria
Roberto Iglesias Rodriguez	Universidad de Santiago de Compostela, Spain
Franziska Kirstein	Blue Ocean Robotics, Denmark
James Little	The University of British Columbia, Canadá

Joaquin Lopez   University of Vigo, Spain
Rebeca Marfil   University of Malaga, Spain
Francisco Martín Rico   Universidad Rey Juan Carlos, Spain
Humberto Martínez Barberá   Universidad de Murcia, Spain
Vicente Matellán   University of León, Spain
Rashid Mehmood   King Abdulaziz University, Saudi Arabia
Manuel Mucientes   Universidade de Santiago de Compostela, Spain
Rafael Muñoz-Salinas   University of Cordoba, Spain
Pedro Núñez   Universidad de Extremadura, Spain
Carlos V. Regueiro   Universidade da Coruña, Spain
Lluís Ribas-Xirgo   Universitat Autònoma de Barcelona, Spain
Humberto Rodríguez   Universidad Tecnológica de Panamá, Panamá
Adrián Romero-Garcés   University of Malaga, Spain
Cristina Romero-González   Universidad de Castilla-La Mancha, Spain
Andrés Rosales   Escuela Politécnica Nacional, Ecuador
Gustavo Scaglia   Universidad Nacional de San Juan, Argentina
Jose Soler   Universitat Politècnica de València, Spain
Nicola Strisciuglio   University of Groningen, Netherlands
Cristina Urdiales   University of Malaga, Spain
Diego Viejo   Universidad de Alicante, Spain
Andrés Vivas   Universidad del Cauca, Colombia
Dimitri Voilmy   Université de Technologie de Troyes, France
Eduardo Zalama   Universidad de Valladolid, Spain

Additional Reviewers

Catalá, Alejandro
Gómez-Donoso, Francisco

Contents

Computer Vision and Robotics

Human Robot Interaction

Mobile Robots

AI and Robotics

Semantic Localization of a Robot in a Real Home

Edmanuel Cruz[1,2]($\boxtimes$), Zuria Bauer[1], José Carlos Rangel[2], Miguel Cazorla[1],
and Francisco Gomez-Donoso[1]

[1] Institute for Computer Research, University of Alicante,
P.O. Box 99, 03080 Alicante, Spain
edcruz@dccia.ua.es
[2] RobotSIS, Universidad Tecnológica de Panamá, Santiago, Panama

Abstract. In social robotics, it is important that a mobile robot knows
where it is because it provides a starting point for other activities such
as moving from one room to another. As a contribution to solving this
problem in the field of the semantic location of the mobile robot, we
pro- pose to implement a methodology of recognition and scene learning
in a real domestic environment. For this purpose, we used images from
five different residences to create a dataset with which the base model
was trained. The effectiveness of the implemented base model is evaluated in different scenarios. When the accuracy of the site identification
decreases, the user provides feedback to the robot so that it can process
the information collected from the new environment and re-identify the
current location. The results obtained reinforce the need to acquire more
knowledge when the environment is not recognizable by the pre-trained
model.

Keywords: Robotics · Deep learning · Semantic localization
CNN training · Neural networks

1 Introduction

It is becoming increasingly common to design and create robots with the ability
to interact with humans, whether it be caring for the disabled or older adults,
or as assistants in shopping malls or receptionists. Social robotics is a research
lines that allows these robots to be easily integrated into their environments. In
this context, the semantic localization of mobile robots plays a crucial role.

Changing between different environments is currently a challenging task
because the systems are unable to adapt to a continually changing environment.
The usual scenario is a system that loses accuracy when the changes become
more drastic. According to the deployment place of the robot the same semantic
categories may also be visually different.

In the same way as a human who first analyzes where he or she is and then
decides where to go or what to do, it is important that the robot recognizes
where it is in order to perform other activities.

© Springer Nature Switzerland AG 2019
R. Fuentetaja Pizán et al. (Eds.): WAF 2018, AISC 855, pp. 3–15, 2019.
https://doi.org/10.1007/978-3-319-99885-5_1

In this work, we propose to implement a methodology of recognition and scene learning in a real domestic environment. The steps carried out are: (1) taking images from five residences; (2) creating a dataset with images from one residence; (3) training a base model; and (4) experimenting with images from the other residences.

The rest of the paper is organized as follows: Sect. 2 presents the state-of-the-art in the field. Then, Sect. 3 details a description of the proposal. This is followed by Sect. 4, where the procedures for testing the proposed approach are described. Next, Sect. 5 details the results of experiments. Finally, Sect. 6 includes the conclusions and future works.

2   Related Works

In recent decades, many researchers have conducted investigations related to semantic localization. We start with works like [1], where the authors train a neural network to estimate the location of a mobile robot in its environment using the odometry information and ultrasound data. The authors in [2] use a pre-programmed routine to detect doorways from range data. In [3], a system was developed with the ability to learn to use a hybrid methodology based on human demonstrations and user advice. Two years later, the authors in [4] describe a virtual sensor that is able to identify rooms from range data. The same year, in [5], the authors apply different learning algorithms to learn topological maps of the environment.

In [6], the authors proposed to use line features to detect corridors and doorways. Furthermore, [7] merges data from local and global features of an image with data from laser sensors. The authors then predict the category of a place by training a Support Vector Machine (SVM) system.

According to [8], many mobile service robots operate in close interaction with humans. They present an approach to people awareness for mobile service robots that utilizes knowledge about the semantics of the environment.

The use of semantic labels was considered in the proposal of [9]. Here, the authors manually insert semantic labels into a 2D image and complement this representation with 3D points.

Before continuing, it is important to mention the tools that have helped the development of semantic localization research. The appearance of the ImageNet Large Scale Visual Recognition Challenge (ILSVRC) [10] allowed algorithms to be evaluated for large-scale object detection and image classification. Researchers were able to compare progress in detection across a wider variety of objects taking advantage of the expensive labeling effort. The work put forward in [11], introduces a scene-centric database called Places with over 7 million labeled pictures of scenes. The authors propose new methods to compare the density and diversity of image datasets and show that the Places dataset is as dense as other scene datasets and has greater diversity. Using CNN, they learn deep features for scene recognition tasks.

Continuing with our review of previous works, developing a robot with a grounded spatial vocabulary is the proposal in [12]. The authors propose a CNN architecture based on engineered features. Such a vocabulary would allow it to give and follow directions and would give it valuable additional information in aiding localization and navigation.

In [13] the authors describe the problem of location as follows. The problem of semantic localization in social robotics could be defined as the identification of the location of a robot by semantic categories representing a place. The traditional approach for solving semantic localization problems is the utilization of semantic categories such as living room or kitchen, together with the robots perceptions as input data for a supervised classification process.

The survey [14] analyzes the different approaches used to address the localization problem. The problem of recognition of changes in environments is also presented in this work. Robust visual localization under a wide range of viewing conditions is a fundamental problem in computer vision. Dealing with the difficulties of this problem is not only highly challenging but also of significant practical importance, e.g., in the context of life-long localization for augmented reality or autonomous robots.

The authors in [15] propose a novel approach based on a joint 3D geometric and semantic understanding of the world, enabling it to succeed under conditions where previous approaches failed.

In [16], the authors propose a method for semantically parsing the 3D point cloud of an entire building using a hierarchical approach.

The authors in [17] provide an overview of indoor localization technologies, popular models for extracting semantics from location data, approaches for associating semantic information and location data, and applications that may be enabled with location semantics. Environment representation for scene classification could be produced by using different kinds of descriptors such as 3D and 2D.

In [18], the authors propose to develop a study for building robust scene descriptors based on the combination of visual and depth data. The approach was tested for classification problems.

The authors in [19] combine semantic web-mining and situated robot perception to develop a system capable of assigning semantic categories to regions of the space.

In [20], the main idea is to leverage the semantic information provided by the user activities and the accurate metric map created by an assistive robot. In [21], the authors proposal includes the evaluation of several CNN classification models in order to find the one that produces the most accurate classification results.

The authors in [22] propose a probabilistic framework that combines human activity sensor data generated by smart wearables with low level localization data generated by robots.

In [23], the authors explore different retraining strategies and experimentation in order to obtain insight about which method provides better precision-training time trade-off. Different settings on the training data are presented, and modifications to different fine-tuning strategies are also explored.

Semantic classification is currently an exciting topic with a great number of published works. In this research, we only focus on current techniques for place categorization that take advantage of DL in order to provide a semantic definition for a place.

3   Proposal

The aim of this study is to implement a methodology to achieve a solid and long-term understanding of the interior scene in changing scenarios. We focus on expanding the work in [23], which presented an optimal methodology for a robot to learn a new environment, from already acquired knowledge. The abovementioned methodology was carried out in a laboratory while the aim of the present work is tested in a real domestic environment.

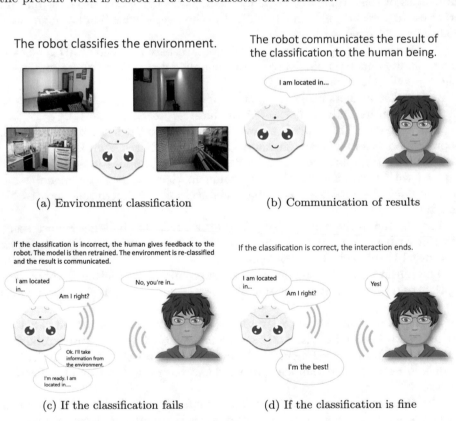

(a) Environment classification   (b) Communication of results

(c) If the classification fails   (d) If the classification is fine

Fig. 1. System interaction

The scenario we wish to present is as follows: first, the robot captures images of the environment and tries to classify them using a previously trained model with images belonging to another place. When the robot is in a new environment, the system is expected to obtain low accuracy due to the differences in the visual features of the new environment. In this case, the user can provide information to the robot to collect data and re-identify the location. If the category provided by the user is not pre-defined, this will be added as a new category, thus allowing the robot to increase its knowledge of the environment. This scenario is shown in Fig. 1.

To complete this goal, we use the neural network architecture shown in Fig. 2. This works as follows: an input image is forwarded to the ResLoc CNN architecture. In this case, we removed the last fully connected layer in order to obtain the visual features descriptor for the input image. As a result, the output of ResLoc CNN is a 2048-dimension feature vector.

The visual features and the respective categories of each image of the dataset are extracted using the ResLoc CNN part of the architecture and inserted into the features database. This feature database is a model that stores the learned data, features of the training samples, and is used during the inference stage.

In the inference stage, the unknown image is forwarded to ResLoc CNN in order to extract the visual feature vector. A K-Nearest Neighbors (KNN) classifier then performs a query on the feature database using the recently computed feature vector. Next, a polling is carried out among the categories of the neighbors, and the most voted category is returned as the final classification of the unknown image.

The performance of the KNN is highly dependent on the k parameter (number of neighbors). Experimentation is carried out using the value of k, which is 3 as it appears in [23]. We used the Annoy[1] implementation of the (approximate) KNN classifier.

4   Experimentation

4.1   Dataset

The experiments described in this work were carried out using our own dataset that provides a semantic category for each RGB image. It is important to clarify that the base model was built with images from only one residence which is identified in the document as House 01. The categories come from the location in which the images were acquired.

Figure 3 shows representative images for the 7 categories available in the dataset.

In order to train the base model we took video sequences from House 01, them randomly shuffled and split them into a 70% training and 30% test ratio. We use only RGB frames.

Table 1 shows the final number of samples per category.

---

[1] https://github.com/spotify/annoy.

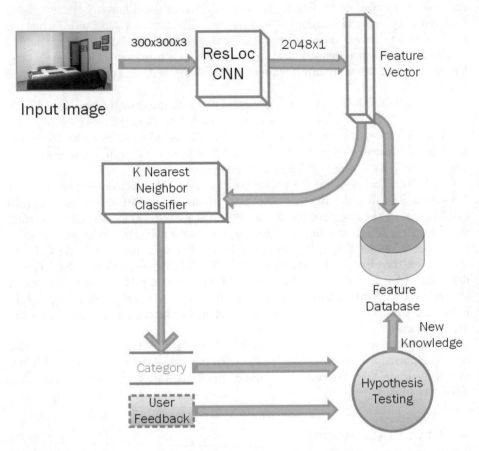

Fig. 2. This Architecture uses the features of a ResLoc CNN with a vector of 2048 features as output. The training samples are forwarded to the ResLoc CNN in order to extract their feature vector. The feature vectors construct the model of a KNN classifier.

Table 1. Images distribution per category.

Cat. ID	Category	Training	Test
1	Corridor	3,782	1,622
2	Dining-living room	4,084	1,751
3	Balcony	1,121	481
4	Kitchen	2,664	1,143
5	Laundry	1,714	735
6	Bathroom	3,932	1,686
7	Bedroom	4,868	2,087

Corridor   Dining-LivingRoom   Balcony

Laundry   Bathroom   Bedroom

Kitchen

Fig. 3. Sample images for each category of our Home Dataset.

4.2 Experimentation Setup

The experiments were carried out as follows:

- Experiments 1 to 2 consist of using a trained model with data from the House 01 in House 02.
- Experiments 3 to 8 consist of using a trained model with data from House 01 in House 03.
- Experiments 9 and 10 consist of using a trained model with data from House 01 in House 04.
- Experiments 11 and 16 consist of using a trained model with data from House 01 in the House 05.

We simulated scenarios in which the robot was incorrectly located in different environments and we obtained feedback from users to correct this knowledge.

For experiments in which new knowledge was included, we used images that were captured in different houses. In these houses, we have the same semantic categories but different visual appearance. The robot then proceeded to capture new information about the environment that the system had failed to identify. Subsequently, the new information was added to the current learned model.

Table 2 shows information on the categories of the different houses. We used this data to validate the efficiency of the model.

Table 2. Images from different houses.

No.	Category	Qty.	House
1	Kitchen	1,113	02
2	Bedroom	1,293	02
3	Bedroom 1	378	03
4	Bedroom 2	330	03
5	Bedroom 3	403	03
6	Bedroom 4	512	03
7	Corridor 1	436	03
8	Corridor 2	529	03
9	Bedroom	359	04
10	Kitchen	212	04
11	Bedroom 1	641	05
12	Bedroom 2	642	05
13	Balcony	571	05
14	Bathroom	427	05
15	Corridor	412	05
16	Kitchen	665	05

5   Results and Discussion

First, we comment on the experiments carried out in the different houses using only the base model, and then we discuss what happened when the system was unsuccessful with the classification and we capture information from the new environment. A summary of the results for the experiments performed can be found in Table 3.

Experiment 1 establishes the baseline we use to compare the following experiments. The total accuracy of the test is 99.98%. This represents the starting line, as no new knowledge was added.

In Experiment 2, a 69.81% success rate was obtained. This was conducted in House 02. This can be considered a considerable success given that we used a completely different environment that the system had never seen before.

Experiment 3 used the bedroom in House 02, obtaining a success rate of 50.27%.

Experiments from 4 to 9 were performed in House 03, obtaining results of (Bedrooms 1 → 48.14%), (Bedrooms 2 → 32.12%), (Bedrooms 3 → 28.53%), (Bedrooms 4 → 60.54%), (Corridor 1 → 85.09%) (Corridor 2 → 75.61%).

As in the kitchen of House 02, the Corridors and Bedroom 4 achieved an acceptable accuracy percentage considering that it was an environment the system had never seen before.

Table 3. Summary of the experiments

No.	Data	House	Acc. without retraining	Acc. with retraining
1	Test	01	99.9	Not applicable
2	Kitchen	02	69.8	100
3	Bedroom	02	50.3	99.5
4	Bedroom 1	03	48.1	100
5	Bedroom 2	03	32.1	100
6	Bedroom 3	03	28.5	100
7	Bedroom 4	03	60.5	100
8	Corridor 1	03	85.1	100
9	Corridor 2	03	75.6	100
10	Bedroom	04	24.5	100
11	Kitchen	04	10.4	100
12	Bedroom 1	05	46.8	100
13	Bedroom 2	05	66.9	100
14	Balcony	05	0.7	100
15	Bathroom	05	48.7	100
16	Corridor	05	44.7	100
17	Kitchen	05	10.5	100

Experiments 10 and 11 were carried out in House 04. Experiment 10 scored 24.51%. This is the lowest success rate obtained in this category. This was mainly due to the visual difference between this category and the images used in the model. Experiment 11 scored 10.37%. As in the previous case, a low percentage was obtained, which was due to the visual difference between this category and the images used in the model.

Experiments 12 to 17 were carried out in House 05, obtaining a result of (Bedrooms 1 → 46.80%), (Bedrooms 2 → 66.97%), (Balcony → 0.70%), (Bathroom → 48.71%), (Corridor → 44.66%) (Kitchen → 10.52%).

It should also be noted that many of the confusions were caused by specific elements appearing in the scenes, such as the case of Experiment 2, which was conducted in a kitchen that had a washing machine, with the system mistaking these images for a laundry (see Fig. 4). When the data was captured, accuracy in all categories increased considerably.

(a) Image from the Laundry category

(b) Image from the Kitchen of House 2

Fig. 4. Image comparison

6   Conclusions and Future Work

As mentioned at the beginning of this study, it is important for the robot to know its location since this is the starting point for other actions it can perform. To provide the robot with the ability to identify its location we used an existing methodology to identify a place. However, this methodology had previously only been developed in the laboratory and had not been tested in a real environment.

After conducting the appropriate evaluations, we conclude that a model will produce better recognition results when tested in an environment similar to that in which it was trained compared to when tested in a different environment.

Our experimental results show that a trained model will obtain more accurate results when re-testing images from the same environment. Furthermore, for images belonging to a different house, the model obtained less accurate results when compared to those for the same house. On the other hand, as was to be expected when adding new knowledge on the environment, the success rate increased considerably.

As future work, we propose to evaluate this approach with different lighting conditions and also introduce more houses in the study. Another feature could be evaluation using 3D information about the places to improve the results using more information.

We also plan to merge the original set of images used in the study with the information generated by user feedback in order to create a full dataset.

Acknowledgements. This work has been supported by the Spanish Government TIN2016-76515R Grant, supported with Feder funds. Edmanuel Cruz is funded by a Panamenian grant for PhD studies IFARHU & SENACYT 270-2016-207. This work has also been supported by a Spanish grant for PhD studies ACIF/2017/243. Thanks to Nvidia also for the generous donation of a Titan Xp and a Quadro P6000.

References

1. Oore, S., Hinton, G., Dudek, G.: A mobile robot that learns its place. Neural Comput. **9**(3), 683–699 (1997)
2. Koenig, S., Simmons, R.: Xavier: a robot navigation architecture based on partially observable Markov decision process models. In: Kortenkamp, D., Bonasso, R., Murphy, R. (eds.) Artificial Intelligence Based Mobile Robotics: Case Studies of Successful Robot Systems, pp. 91–122. MIT-Press, Cambridge (1998)
3. Dillmann, R., Rogalla, O., Ehrenmann, M., Zöliner, R., Bordegoni, M.: Learning robot behaviour and skills based on human demonstration and advice: the machine learning paradigm. In: Hollerbach, J.M., Koditschek, D.E. (eds.) Robotics Research, pp. 229–238. Springer, London (2000)
4. Buschka, P., Saffiotti, A.: A virtual sensor for room detection. In: Proceedings of the IEEE/RSJ International Conference on Intelligent Robots and Systems (IROS), pp. 637–642 (2002)
5. Kuipers, B., Beeson, P.: Bootstrap learning for place recognition. In: Proceedings of the National Conference on Artificial Intelligence (AAAI) (2002)
6. Althaus, P., Christensen, H.: Behaviour coordination in structured environments. Adv. Robot. **17**(7), 657–674 (2003)
7. Pronobis, A., Mozos, O.M., Caputo, B., Jensfelt, P.: Multi-modal semantic place classification. Int. J. Robot. Res. **29**(2–3), 298–320 (2010). https://doi.org/10.1177/0278364909356483
8. Stückler, J., Behnke, S.: Improving people awareness of service robots by semantic scene knowledge. In: Ruiz-del-Solar, J., Chown, E., Plöger, P.G. (eds.) RoboCup 2010: Robot Soccer World Cup XIV, pp. 157–168. Springer, Heidelberg (2011). ISBN: 978-3-642-20217-9

9. Lim, H., Sinha, S.: Towards real-time semantic localization. In: ICRA Workshop on Semantic Perception (2012)
10. Russakovsky, O., Deng, J., Su, H., Krause, J., Satheesh, S., Ma, S., Huang, Z., Karpathy, A., Khosla, A., Bernstein, M., Berg, A.C., Fei-Fei, L.: ImageNet large scale visual recognition challenge. IJCV **115**, 211–252 (2015)
11. Zhou, B., Lapedriza, A., Xiao, J., Torralba, A., Oliva, A.: Learning deep features for scene recognition using places database. In: Ghahramani, Z., Welling, M., Cortes, C., Lawrence, N.D., Weinberger, K.Q. (eds.) Advances in Neural Information Processing Systems 27, pp. 487–495. Curran Associates, Inc. (2014), http://papers.nips.cc/paper/5349-learning-deep-features-for-scene-recognition-using-places-database.pdf
12. Goeddel, R., Olson, E.: Learning semantic place labels from occupancy grids using CNNs. In: 2016 IEEE/RSJ International Conference on Intelligent Robots and Systems (IROS), pp. 3999–4004, October 2016
13. Martinez-Gomez, J., Gimenez, V.M., Cazorla, M., Garcia-Varea, I.: Semantic Localization in the PCL library. Robot. Auton. Syst. **75**(Part B), 641–648 (2016). http://www.sciencedirect.com/science/article/pii/S0921889015001943
14. Lowry, S., Sünderhauf, N., Newman, P., Leonard, J.J., Cox, D., Corke, P., Milford, M.J.: Visual place recognition: a survey. IEEE Trans. Robot. **32**(1), 1–19 (2016). https://doi.org/10.1109/TRO.2015.2496823
15. Schünberger, J.L., Pollefeys, M., Geiger, A., Sattler, T.: Semantic visual localization. CoRR, abs/1712.05773,2017. http://arxiv.org/abs/1712.05773
16. Armeni, I., Sener, O., Zamir, A.R., Jiang, H., Brilakis, I., Fischer, M., Savarese, S.: 3D semantic parsing of large-scale indoor spaces. In: The IEEE Conference on Computer Vision and Pattern Recognition (CVPR), June 2016
17. Ma, S., Liu, Q.: Semantic localization. In: Encyclopedia with Semantic Computing and Robotic Intelligence, vol. 01, no. 01 (2017). https://doi.org/10.1142/S242503841630010X
18. Romero-González, C., Martínez-Gómez, J., García-Varea, I., Rodríguez-Ruiz, L.: On robot indoor scene classification based on descriptor quality and efficiency. Expert Syst. Appl. **79**, 181–193 (2017). http://www.sciencedirect.com/science/article/pii/S0957417417301318
19. Young, J., et al.: Making sense of indoor spaces using semantic web mining and situated robot perception. In: Blomqvist, E., Hose, K., Paulheim, H., Lawrynowicz, A., Ciravegna, F., Hartig, O. (eds.) The Semantic Web: ESWC 2017 Satellite Events. ESWC 2014. Lecture Notes in Computer Science, vol. 10577. Springer, Cham (2017). https://doi.org/10.1007/978-3-319-70407-439
20. Rosa, S., Lu, X., Wen, H., Trigoni, N.: Leveraging user activities and mobile robots for semantic mapping and user localization. In: Proceedings of the Companion of the 2017 ACM/IEEE International Conference on Human-Robot Interaction (HRI 2017), pp. 267–268. ACM, New York (2017). https://doi.org/10.1145/3029798.3038343
21. Cruz, E., Rangel, J.C., Cazorla, M.: Robot semantic localization through CNN descriptors. In: Ollero, A., Sanfeliu, A., Montano, L., Lau, N., Cardeira, C. (eds.) ROBOT 2017: Third Iberian Robotics Conference, pp. 567–578. Springer, Cham (2018). https://doi.org/10.1145/3215525.3215526

22. Rosa, S., Patané, A., Lu, X., Trigoni, N.: CommonSense: collaborative learning of scene semantics y robots and humans. In: IoPARTS 2018: 1st International Workshop on Internet of People, Assistive Robots and ThingS, Munich, Germany, 10 June 2018, 6 pages. ACM, New York (2018). https://doi.org/10.1145/3215525.3215526

23. Cruz, E., Rangel, J.C., Gómez-Donoso, F., Bauer, Z., Cazorla, M., Garcia-Rodriguez, J.: Finding the place: how to train and use convolutional neural networks for a dynamically learning robot. In: 2018 International Joint Conference on Neural Networks, IJCNN, pp. 3655–3662, July 2018. ISBN: 978-1-5090-6014-6

Positioning System for an Electric Autonomous Vehicle Based on the Fusion of Multi-GNSS RTK and Odometry by Using an Extented Kalman Filter

Miguel Tradacete[✉], Álvaro Sáez[✉], Juan Felipe Arango[✉],
Carlos Gómez Huélamo[✉], Pedro Revenga[✉], Rafael Barea[✉],
Elena López-Guillén[✉], and Luis Miguel Bergasa[✉]

Robesafe Group, Department of Electronics, University of Alcala,
Alcalá de Henares, Spain
tradacete.miguel@gmail.com,
{alvaro.saezc,juanfelipe.arango,carlos.gomezh}@edu.uah.es,
{rafael.barea,elena.lopezg}@uah.es
https://www.robesafe.com/

Abstract. This paper presents a global positioning system for an autonomous electric vehicle based on a Real-Time Kinematic Global Navigation Satellite System (RTK- GNSS), and an incremental-encoder odometry system. Both elements are fused to a single system by an Extended Kalman Filter (EKF), reaching centimeter accuracy. Some varied experiments have been carried out in a real urban environment to compare the performance of this positioning architectures separately and fused together. The achieved aim was to provide autonomous vehicles with centimeter precision on geolocalization to navigate through a real lane net.

Keywords: Autonomous vehicle · Positioning · Odometry
Multi-GNSS · Kalman Filter

1 Introduction

Vehicle positioning and tracking have numerous applications in general transport-related studies including vehicle navigation, fleet monitoring, traffic congestion etc. In the last decade, many works have been focused in studying driving behaviour through examining the vehicle movement trajectory using GNSS signals, mostly GPS [1–3]. These methods have been able to provide both geolocalization and time information to a receiver employing multiple satellite signals while they stay fast, accurate, and cost-efficient. However, their performance has a strong dependence on several system factors such as the number of visible satellites, their positions or the capability of the GPS receiver. In addition, the signal trips through the layers of the atmosphere, and some other

© Springer Nature Switzerland AG 2019
R. Fuentetaja Pizán et al. (Eds.): WAF 2018, AISC 855, pp. 16–30, 2019.
https://doi.org/10.1007/978-3-319-99885-5_2

sources contribute to inaccuracies and errors in the GPS signals by the time they reach a receiver. Thus, the accuracy provided by this methods is low (usually between 1 and 10 m), they need a considerable time (over 30 s) to provide the first position measurement and they do not guarantee a robust service in several situations such as environments with poor signal conditions.

The development of Intelligent Vehicles (IVs) has specially grown during the last years. These systems aim to solve complex issues with specially demanding accuracy requirements (usually decimeter precision) like autonomous driving applications where tasks such as lane maintenance analysis demand centimeter precision [4]. Furthermore, autonomous vehicles also require robust solutions with low latencies and high time availability so standalone GNSS techniques are not adequate for them.

Various solutions are proposed to achieve a better service quality: to deal with the accuracy problem Differential GPS (DGPS) is used to obtain an accuracy enhancement using data from a reference station [5,6] and the more complex Real-time Kinematic (RTK) positioning solution, which uses carrier phase information, has attracted much interest in applications with strict precision requirements due to its centimeter-level accuracy [7]. To approach the robustness issue Multi-GNSS (multiple Global Navigation Satellite Systems) techniques are being widely-used, boosted by the appearance of alternative GNSSs based on different satellite constellations like Russian (GLONASS), European Union (Galileo), Chinese (Beidou) or Japanese (QZSS). Multi-GNSS allows to easily increase the number of tracked satellites to over 10 in good signal conditions and to more than 5 in almost any other situation, even including dense urban areas combining multiple GNSS [4]. Several studies have proven the benefits of these techniques combining GPS and GLONASS [8,9], GPS and Galileo [10] or even four of the available systems (GPS+Galileo+BeiDou+QZSS) [11].

Nevertheless, even the combination of the previous methods might not be enough to cover autonomous vehicles needs in certain environments such as dense urban or concrete places like tunnels. To face this challenging situations, GNSS data needs to be fused with local sensors information when the measurement's quality is degraded. In [12] RTK-DGPS was fused with speed vehicle sensors and steering-wheel position measurements to improve vehicle tracking. Other works like [13] used an Extended Kalman Filter to integrate DGPS with some vehicle sensors like an inertial navigation system (INS) through a kinematic model in order to achieve enough accuracy to enable vehicle cooperative collision warning without the use of ranging sensors.

This paper presents a robust real-time positioning system for autonomous vehicles that reaches centimeter precision. The system uses a GNSS receiver and an incremental-encoder odometry, integrated by an Extended Kalman Filter which leverages quality of the received satellite measurements. As well as, odometry system is calibrated through an automatic process applying a least square adjustment of the position error of a variety of routes. Experiments presented in Sect. 4 show that our system is able to keep the vehicle in the middle of the lane nets even in regions without available differential corrections.

Furthermore, the system is complemented with a reactive navigation module based on vision and Lidar that slightly relaxes the positioning requirements.

This paper is organized as follows: Sect. 2 presents the system's structure together with an analysis of the main modules that compose it and their corresponding standalone performances. Section 3 analyzes the integration of both modules using the Extended Kalman Filter and the following Sect. 4 exposes the results of the performed experiments to test the final system with different configurations. Last Sect. 5 presents the final conclusions and future work lines.

2   System Architecture

The positioning system is integrated in an open-source electric vehicle (TABBY EVO Vehicle 4 seats version) modified and automatized by the University of Alcalá. The system's architecture includes a Real-Time Kinematic Multi-Global Navigation Satellite System based on both GPS and GLONASS with a local base station that broadcasts differential corrections, a GPRS modem, and an incremental-encoder odometry system. Its sensor equipment is composed of a GNSS receiver, a Choke-Ring Antenna for the local base station, and two Kübler 3700 incremental encoders for odometry. All these modules are fused in Robotic Operating System (ROS) using an Extended Kalman Filter. Figure 1 shows the general diagram of the system.

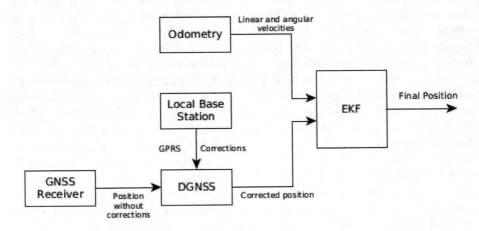

Fig. 1. System architecture diagram

GNSS receiver is set on top of the vehicle to obtain maximum coverage, and Choke-Ring Antenna for a base station is set on the Polytechnic School building's roof. The odometry encoders are assembled in both rear-wheel shafts by 3D-printed pieces. ROS runs on two embedded GPUs looking for the benefits of modularity. These GPUs are a Nvidia Jetson TX2, and a Raspberry Pi 3 for odometry processing. Figure 2 displays the entire vehicle, and Fig. 3(a) (where

Fig. 2. General view of the TABY EVO-OSVehicle

(a) GPS and Lidar   (b) Incremental encoder

Fig. 3. Vehicle sensor equipment

Lidar is shown as part of reactive navigation system) and Fig. 3(b) show GNSS receiver and odometry encoder details.

2.1 Multi-GNSS

The main module of the localization system consists of a multi-constellation system (multi-GNSS) with RTK positioning solution. In addition, it also includes two elements: a differential Hiper Pro GPS+ receiver configured as rover, and a local base station to generate differential corrections.

The rover is able to obtain data from both GLONASS and GPS to provide a more robust solution than a standard GPS by increasing the number of visible satellites. It provides positioning information at 10 Hz as autonomous vehicles demand real-time information. Furthermore, it uses differential corrections to improve the achieved accuracy.

The Standard National Geographic Institute of Spain (IGN) public reference station's frequency is usually below 1 Hz which is inadequate for kinematic applications. In addition, the published data only allows positioning sub-meter level accuracies when the rover is within a short distance from the station. To achieve the needed requirements a local base station was deployed on the Polytechnic School building's roof.

The local base station uses a Choke-Ring Antenna, specifically chosen to deal with multipath. The Antenna is connected to a second Hiper Pro GPS + receiver which generates the differential corrections. Corrections are generated at 10 Hz, published over the Internet using standard Open Source software, and then acquired by the car using a GPRS modem linked via radio.

The GNSS module (base station + rover) was tested separately to evaluate the accuracy it was able to provide. In the first trial, the receiver was tested on standalone mode without adding corrections. To perform the trial an hour of data was recorded on another base station with a known position. The results were then compared with the real station position. Figure 4 shows the results of the experiment. The first graphic, Fig. 4(a), represents the deviation of the measurements from the real antenna position in meters. The second Fig. 4(b) represents the RMS error of the measurements (m) as a function of the GPS time.

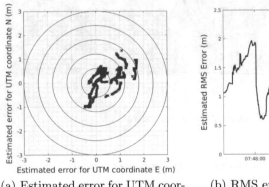
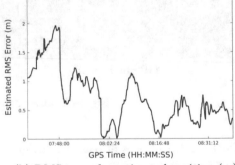

(a) Estimated error for UTM coordinates (m)   (b) RMS error for estimated position (m)

Fig. 4. GPS+GLONASS performance for standalone receiver

The presented results clearly show that standalone system is not enough for autonomous vehicle applications. With a mean measurement value of 71 cm, the graphs expose the system only provides sub-meter accuracy 72% of the time and for 95% it can only ensure an accuracy of 1.7 m. Besides, the tested system clearly shows a considerable lack of repeatability when it is analyzed from one day to another.

In a second trial, differential corrections were added to the receiver and the first test was repeated the same day with similar conditions. Another hour of data was recorded after the ambiguity resolution process, in the receiver, and was completed in order to achieve the best accuracy with the Fixed-RTK solution.

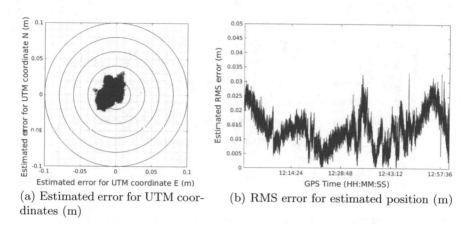

(a) Estimated error for UTM coordinates (m)

(b) RMS error for estimated position (m)

Fig. 5. GPS+GLONASS performance for differential receiver with Fixed-RTK

Figure 5 shows the enhancement that corrections offer: the mean measurement error value is reduced to 1.3 cm with a max deviation of 3.4 cm. The system guarantees centimeter precision 100% of the time and it becomes repeatable if tested on independent days.

Ambiguity resolution process (required to fix the number of wavelengths between satellites and base station, and therefore it provides the optimum solution) is another important issue in the system's performance. Without a fixed solution the accuracy is clearly downgraded. Achieving a fixed-RTK solution takes a long time and it may even be unreachable due to poor satellite visibility. Figure 6 presents the required time to achieve the optimum solution based on collected data during a period of time of two months in favorable conditions (open sky).

The results in Fig. 6 show that the mean required time (50%) to achieve convergence is over 9 min. Furthermore, about 10% of the measurements needed more than 15 min to reach centimeter accuracy even with good satellite signal conditions. The convergence time highlights the need to use complementary systems that improve the main module precision when a Fixed solution is not available.

2.2   Incremental-Encoder Odometry

The implemented odometry system is based on incremental encoders which measure the rear wheels' rotation. Each wheel has its own shaft, which allows the

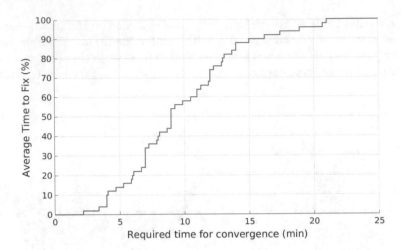

Fig. 6. Convergence time for RTK-Fixed solution.

rotation angle of the vehicle to be measured. This odometry system provides the angular and linear velocity of the vehicle to the Extended Kalman Filter.

The relative position and angle are not given to the filter because of cumulative error. Some external causes, such as the irregularity of the road, add error to the measurement each sampling period. The cumulative error is removed by giving instant velocities.

This odometry system is composed of two main modules: a real-time pulse counter, and an algorithm processor. The capturing of pulses is a critical task since missing them implicates losing control of the measurement, and a systematic error. An Olimexino-STM32 is in charge of the counter. Then, the calculations of lineal and angular velocities are performed by a Raspberry Pi 3. This task uses the time between each interaction given by the pulse counter to carry out the calculations of the velocities. These two modules communicate each other through serial-communication protocol.

The calculations of lineal and angular velocities are referenced to the central point between rear wheels, following an Ackerman model [14] as shown in Fig. 7.

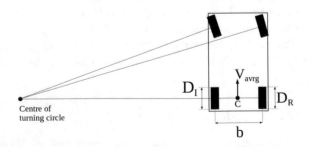

Fig. 7. Vehicle diagram - Ackerman steering geometry

The lineal velocity V_{avrg} is obtained by Eq. 1 as an average of the two lineal velocities of the wheels, where as V_R is the velocity of the right wheel and V_L is the velocity of the left.

$$V_{avrg} = \frac{V_R + V_L}{2} \tag{1}$$

Both independent velocities of each wheel are calculated as

$$V_R = \frac{N_R D_R \pi}{PT} \tag{2}$$

and

$$V_L = \frac{N_L D_L \pi}{PT} \tag{3}$$

where.

N_R number of pulses of the right wheel
N_L number of pulses of the left wheel
D_R diameter of the right wheel
D_L diameter of the left wheel
P encoder resolution (in pulses per revolution of the wheel)
T time between each interaction of calculations

Angular velocity ω_{avrg} is obtained by Eq. 4 as a lineal derivation of the developed rotation angle θ in each interaction of calculations. This angle is calculated as Eq. 5.

$$\omega_{avrg} = \frac{\theta}{T} \tag{4}$$

$$\theta = \frac{(N_R D_R - N_L D_L)\pi}{bP} \tag{5}$$

where b is the distance between the rear wheels.

The angle is calculated in this manner as a consequence of the fact the model of the vehicle is based on the Ackerman steering geometry (Fig. 7), which implies that every movement is a circular curve. Although this model is only kinematic, it is a appropriate for low velocities [15].

To calibrate the odometry parameters we implemented an automatic process that analyzes different routes to eliminate systematic errors in the calculated position. These errors come from variations in the diameter of the wheels, and the real distance between the wheels. In the performed routes, the position tracks are analyzed, and therefore the calibration process is able to obtain the real odometry parameters D_R, D_L, and b using a least squares adjustment of the position error at the end of the routes.

In order to calibrate the parameters, first, a straight movement has to be analyzed to obtain the real dimensions of the wheels to find the minimum error. Figure 8(a) shows the surface of error in one test, which has a minimum at $D_R = 560.4\,\text{mm}$ and $D_L = 558\,\text{mm}$. The reasoning behind using a straight route comes from the fact that no turn is made, meaning the distance between wheels b does not interact.

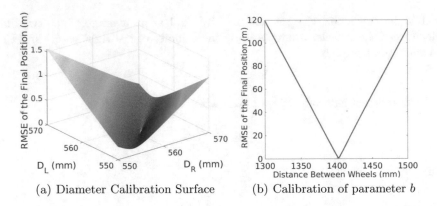

(a) Diameter Calibration Surface   (b) Calibration of parameter b

Fig. 8. Odometry calibration

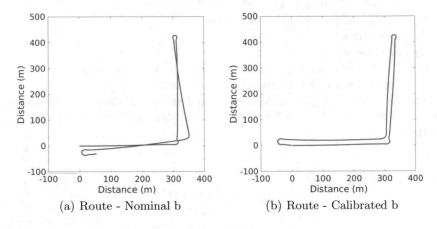

(a) Route - Nominal b   (b) Route - Calibrated b

Fig. 9. Complex route results

Then, using the calculated dimensions of the wheels, a complex route is analyzed to establish the distance between wheels. The results of the final position
error are shown in Fig. 8(b), with a minimum at $b = 1404$ mm. Finally, Fig. 9(a)
shows the complex route using the nominal value of b (1350 mm), and Fig. 9(b)
shows the same route using the calibrated value of b (1404 mm).

3   Extended Kalman Filter

Standard EKF algorithm and formulation are widely known [16]. To obtain the
output of EKF measurement covariance matrices associated with localization
sensors are employed. Originally, the raw covariance matrices provided by the
sensors were used. However, these matrices only considered certain quality indicators, which were insufficient for the application requirements. For example,
the used GNSS receiver only considers some parameters (such as HDOP) but

does not leverage the usage of differential corrections, which provide reliable information about measurement accuracy.

Thus, we propose an adaptive filter in accordance with every available quality parameter in order to adapt the output of the filter to the real environment conditions.

3.1   Algorithm

The developed EKF aims to estimate the full 3D pose and velocity of the vehicle over time using the information provided by the previously discussed sensors. The vehicle's system state is a six-element vector that comprises the vehicle's 3D orientation and velocity. It is calculated as Eq. 6 where f represents nonlinear state transition function and w_{k-1} is the process noise.

$$x_k = f(x_{k-1}) + w_{k-1} \tag{6}$$

In addition, each employed sensor provides measurements that are modeled as 7:

$$z_k = h(x_k) + v_k \tag{7}$$

where h is a nonlinear state transition sensor function and v_k is the measurement's noise which is assumed to be normally distributed.

The prediction stage is described by Eqs. 8 and 9 where a standard 3D kinematic model as product of Newtonian mechanics is used as f. F is the Jacobian of f which is used to project the covariance error P, and finally, Q is the processed noise covariance.

$$\hat{x} = f(x_{k-1}) \tag{8}$$
$$\hat{P}_k = F P_{k-1} P^T + Q \tag{9}$$

The correction stage is carried out through Eqs. 10, 11 and 12. The first one calculates the Kalman gain using the sensors measurement matrix H, the measurement covariance, and the estimated error covariance $\hat{P}_k$. The gain is employed to update the final state vector and the covariance matrix.

$$K = \hat{P}_k H^T (H \hat{P}_k H^T + R)^{-1} \tag{10}$$
$$x_k = \hat{x_k} + K(z - H\hat{x_k}) \tag{11}$$
$$P_k = (I - KH)\hat{P}_k(I - KH)^T + KRK^T \tag{12}$$

The measurement covariance matrices provided by the sensors are adjusted following the GNSS receiver quality indicators: fix quality and Horizontal Dilution of Precision (HDOP). During the first step, the covariance matrices are modified according to fix quality. For a fix quality of 4, the main sensor achieves centimeter precision so matrices are configured to strongly prioritize its information. However, with fix quality of 5, the GNSS receiver has not completed ambiguity resolution process (obtained accuracy is sub-metric), therefore the covariance matrices are adjusted using a linear function dependent on HDOP. Finally, for a fix quality of 1, the GNSS receiver covariance matrix is penalized in benefit of odometry. As it does not provide accurate data, that matrix is adjusted with a linear function as fix quality 5 case.

4   Results

To test the localization system, an area around the UAH campus in Madrid was chosen with approximately 4 Km radius distance to the differential corrections base-station, enabling high quality corrections. The test route presented a length of about 5,5 km with both two lane and four lane roads involving challenging tasks for autonomous vehicles such as pedestrian crossings, roundabouts, and stop and give way signs. Additionally, the route has been performed between 20 to 30 km/h, including defiant conditions for the GNSS signals (trees, high buildings, street signing, electrical lines, etc.).

Below we present results of a sample of stretches in the tested routes, showing different configurations and the performance of the system dealing with real-life, challenging situations. In all the figures, the blue trail is the route determined by the GNSS receiver, the yellow one is the path determined by odometry, and the red one is the EKF output.

Figure 10 presents the data collected by sensors along the main section of the route. GNSS receiver achieved the RTK Fix solution during most of the path, but corrections were lost several times due to multipath. The exposed odometry data did not include a calibration process, as it can clearly been observed, but even without the adjustment the output of the system still acts robustly, even under unfavorable conditions. Figure 11 shows the adaptive EKF output for the previous route.

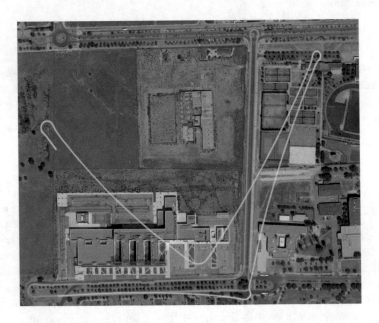

Fig. 10. GNSS standalone fix 1 (blue) and Odometry (yellow)

Fig. 11. Adapted EKF output

Figure 12 presents different details of the sections (output EKF in red and GNSS output in blue dots) where GNSS signals are degraded by multipath faced by three different system configurations. Figure 12(a) demonstrates the performance of the positioning system without the adaptive Kalman. The results clearly show that basic configuration (with sensor raw covariance matrices) strongly depends on the main module performance and fails when GNSS signals are degraded or lost. Figure 12(b) shows the same section for an adaptive Kalman without a RTK solution available. The performance of the system is clearly improved, even in worse GNSS signal conditions, however the system is still not able to maintain the vehicle between lanes. Figure 12(c) presents the performance of the final system with the adaptive Kalman and a RTK positioning solution, which responds correctly to GNSS signal quality degradation.

Figure 13(a) shows the output of the system in a roundabout for a standalone GNSS solution with calibrated odometry and Fig. 13(b) shows EKF output (red trail) just using odometry information. The performance of the odometry is penalized by sharp turns, so roundabouts are challenging environments for the positioning system if the main module is not able to achieve centimeter accuracy. However, the presented results display the capability of the system to respond properly to this challenging condition.

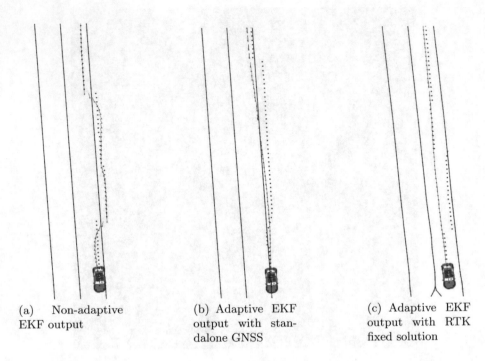

(a)   Non-adaptive EKF output

(b) Adaptive EKF output with standalone GNSS

(c) Adaptive EKF output with RTK fixed solution

Fig. 12. Adaptive EKF robustness example

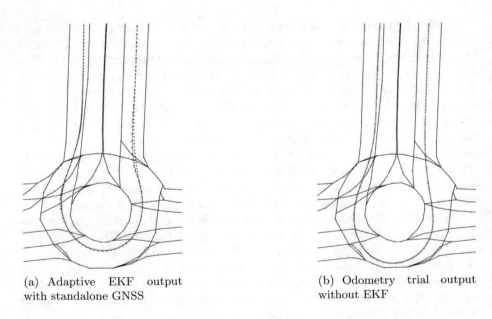

(a) Adaptive EKF output with standalone GNSS

(b) Odometry trial output without EKF

Fig. 13. Improvement over standalone odometry configuration

5   Conclusion

This paper presents an accurate and robust positioning system specifically designed for a fully autonomous vehicle which achieves centimeter precision, even in disadvantageous environments. This positioning system is finally used by a real autonomous vehicle to drive through a lane net, and consequently, in a real urban environment.

The exhibited tests show various potential configurations for the system, which demonstrate the need to use the proposed adaptive EKF in order to correctly handle poor sensor performance situations. The importance of using this adaptive EKF lies in the fact that lanes are just a couple meters width. This means, while using standalone GNSS (fix 1 quality) with an non-adaptive EKF, that one meter of error in GNSS receiver positioning could cause a collision.

Future work involves the integration of additional sensors such as a compass and some IMUs to detect possible skids at high velocities, the adaptation of the EKF to more defined, alternative environments such as dense, urban ones, and the readjust of covariance matrix adaptation of all sensors considering the new ones (compass and IMUs). In addition to this, to improve the odometry in real time while driving, the calibration process will be implemented on board using a variety of stretches as test routes when GNSS receiver has Fixed quality.

Acknowledgment. This work has been partially funded by the Spanish MINECO/FEDER through the SmartElderlyCar project (TRA2015-70501-C2-1-R), the DGT through the SERMON project (SPIP2017-02305), and from the RoboCity2030-III-CM project (Robótica aplicada a la mejora de la calidad de vida de los ciudadanos, fase III; S2013/MIT-2748), funded by Programas de actividades I+D (CAM) and cofunded by EU Structural Funds.

References

1. Greaves, S., Somers, A.: Insights on driver behaviour: what can global positioning system (GPS) data tell us? Publication of ARRB Transport Research, Limited (2003)
2. Ren, H., Xu, T., Li, X.: Driving behavior analysis based on trajectory data collected with vehicle-mounted GPS receivers. Wuhan Daxue Xuebao (Xinxi Kexue Ban)/Geomatics and Information Science of Wuhan University, vol. 39, no. 6, pp. 739–744 (2014)
3. Wong, I.Y.: Using GPS and accelerometry to assess older adults' driving behaviours and performance: challenges and future directions (2013)
4. Sun, Q.C., Odolinski, R., Xia, J.C., Foster, J., Falkmer, T., Lee, H.: Validating the efficacy of GPS tracking vehicle movement for driving behaviour assessment. Travel. Behav. Soc. **6**, 32–43 (2017)
5. Gao, Y., Li, Z., McLellan, J.: Carrier phase based regional area differential GPS for decimeter-level positioning and navigation. In: Proceedings of the 10th International Technical Meeting of the Satellite Division of The Institute of Navigation (ION GPS 1997), pp. 1305–1313 (1997)
6. Ragheb, A.E., Ragab, A.F.: Enhancement of GPS single point positioning accuracy using referenced network stations. World Appl. Sci. J. **18**(10), 1463–1474 (2012)

7. Thitipatanapong, R., Wuttimanop, P., Chantranuwathana, S., Klongnaivai, S., Boonporm, P., Noomwongs, N.: Vehicle safety monitoring system with next generation satellite navigation: Part 1 lateral acceleration estimation. Technical report, SAE Technical Paper (2015)
8. Alkan, R.M., İlçi, V., Ozulu, I.M., Saka, M.H.: A comparative study for accuracy assessment of PPP technique USING GPS and GLONASS in urban areas. Measurement **69**, 1–8 (2015)
9. Angrisano, A., Gaglione, S., Gioia, C.: Performance assessment of GPS/GLONASS single point positioning in an urban environment. Acta Geodaetica et Geophysica **48**(2), 149–161 (2013)
10. Verhagen, S., Odijk, D., Teunissen, P.J., Huisman, L.: Performance improvement with low-cost multi-GNSS receivers. In: 2010 5th ESA Workshop on Satellite Navigation Technologies and European Workshop on GNSS Signals and Signal Processing (NAVITEC), pp. 1–8. IEEE (2010)
11. Odolinski, R., Teunissen, P.J., Odijk, D.: Combined BDS, galileo, QZSS and GPS single-frequency RTK. GPS Solut. **19**(1), 151–163 (2015)
12. Naranjo, J.E., González, C., García, R., De Pedro, T., Haber, R.E.: Power-steering control architecture for automatic driving. IEEE Trans. Intell. Transp. Syst. **6**(4), 406–415 (2005)
13. Rezaei, S., Sengupta, R.: Kalman filter-based integration of DGPS and vehicle sensors for localization. IEEE Trans. Control. Syst. Technol. **15**(6), 1080–1088 (2007)
14. Weinstein, A.J., Moore, K.L.: Pose estimation of Ackerman steering vehicles for outdoors autonomous navigation. In: 2010 IEEE International Conference on Industrial Technology (ICIT), pp. 579–584. IEEE (2010)
15. LaValle, S.M.: Planning Algorithms. Cambridge University Press, Cambridge (2006)
16. Welch, G., Bishop, G.: An introduction to the Kalman filter. University of North Carolina, Department of Computer Science. Technical report, TR 95-041 (1995)

Self-driving a Car in Simulation Through a CNN

Javier del Egio[✉], Luis Miguel Bergasa[✉], Eduardo Romera[✉],
Carlos Gómez Huélamo[✉], Javier Araluce[✉], and Rafael Barea[✉]

Robesafe Group, Department of Electronics, University of Alcalá,
Alcalá de Henares, Spain
{javier.egido,eduardo.romera,carlos.gomezh,javier.araluce}edu.uah.es,
{luism.bergasa,rafael.barea}@uah.es

Abstract. This work presents a comparison between different Convolutional Neural Network models, testing its performance when it leads a self-driving car in a simulated environment. To do so, driving data has been obtained manually driving the simulator as ground truth and different network models with diverse complexity levels has been created and trained with the data previously obtained using end-to-end deep learning techniques. Once this CNNs are trained, they are tested in the driving simulator, checking their ability of minimizing the car distance to the center of the lane, its heading error and its RMSE. The neural networks will be evaluated according to these parameters. Finally, conclusions will be drawn about the performance of the different models according to the parameters mentioned before in order to find the optimum CNN for the developed application.

Keywords: Convolutional Neural Network (CNN) · Self-driving

1 Introduction. State of the Art

Transport has been a fundamental pillar of the economical evolution history since the wheel invention was made five thousand years ago. Many improvements has been made through all these years, reaching the current vehicle which allows people and goods to cover long distances without effort.

Nevertheless, many lives are lost every year due to car accidents, which are mainly produced by human factors such as fatigue, distractions or imprudences.

To solve these problems, driving aids have been implemented in new cars. These systems alerts the driver when the vehicle crosses road lines or automatically brakes the vehicle when an obstacle is on the road.

Depending on technologies implemented in the vehicle, there are different self-driving levels according to Society of Automotive Engineers (SAE) [1]:

- Level 1. Without assistance. Traditional vehicle controlled by a human driver.
- Level 2. The vehicle can control longitudinal or lateral movements under certain conditions.

R. Fuentetaja Pizán et al. (Eds.): WAF 2018, AISC 855, pp. 31–43, 2019.
https://doi.org/10.1007/978-3-319-99885-5_3

– Level 3. The vehicle can control longitudinal and lateral movements but cannot detect eventualities, so the human driver must stay in alert.
– Level 4. The vehicle does not need a human driver. It can move autonomously and acts when system failures are detected, but it only works under certain conditions.
– Level 5. The vehicle does not need a human driver. It can move autonomously under all conditions.

Nowadays the car is quickly evolving into autonomous machines with the ability of interact with the environment and solve driving situations without any problem, being able to drive following the rules as a human but avoiding the problems mentioned above.

1.1 Autonomous Systems

There are two main approaches to control an autonomous car, which will be introduced next.

Traditional Perception and Navigation Techniques [2]. The traditional way to control a self-driving vehicle is to implement a lot of code for perception techniques distinguishing between different classes of objects seen by diverse technologies such as RGB cameras, LIDAR, radar. This information feeds traditional navigation algorithms based on mapping, planning, low-level control and high-level control in charge with the decision making maneuvers. The methods need a big process capacity and expensive technologies that increase the costs.

End-to-End Learning by Using Neural Networks [3]. The incipient way studied in this project, much more simpler, feeds a deep convolutional neural network which controls the car movements with just simple images and applying an end-to-end strategy. The code needed to implement this technology is easier to implement. The used neural network learns how to drive by taking information about manual driving by a human.

This paper researches into Artificial Neural Networks, and more specifically, into Convolutional Neural Networks and its architecture, to develop a new CNN with the capacity to control lateral and longitudinal movements of a vehicle in an open-source driving simulator replicating the human driving behavior.

1.2 Previous Developments

As mentioned before, End-to-End learning is the new way to control autonomous cars. Due to this, many recent tools have been developed to be able to train and test the CNNs ability to control an autonomous car. Some of them are mentioned next.

NVIDIA Self-driving Car [3]. This project trains a Convolutional Neural Network to analyze a single front-facing camera to control steering commands of a real car, monitoring how many times the driver has to take control of the car.

Europilot [4]. Europilot is a platform that allows to control Euro Truck Simulator 2 (a driving simulator game) with an Artificial Neural Network programmed in Keras or Tensorflow.

Gazebo [5]. Gazebo is a realistic city simulator created to train neural networks to control autonomous cars.

Udacity Self-Driving Car Simulator [6]. This project, produced by Udacity, provides a racing driving simulator and tools to communicate the car and the Convolutional Neural Network.

1.3   Training the CNNs

Training data are collected by driving a car in an open-source driving simulator by a human, obtaining images from a front-facing camera and synchronizing steering angle and throttle values performed by the driver. Steering angle values are relative to the maximum (r/r_{max}), where r_{max} means the maximum steering value (25), oscillating between -1 when turning left and 1 when turning right, and throttle values oscillates between 1 (maximum acceleration) and -1 (maximum braking).

The dataset is augmented by horizontal flipping, changing steering angles sign and taking information from left and right cameras from the car, using them as a center image by modifying the steering value with a deviation factor, established as 0.2. Finally, training data is used in the same order as it was taken to correctly train the LSTM layers (Fig. 1).

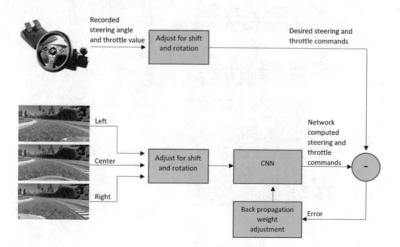

Fig. 1. Left, center and right images from driving simulator captured to train CNNs

When the network is being trained, single images are provided to the CNN as input, obtaining output values which are minimized regarding to output values from dataset created by a human driver through multiple epochs.

2   Network Architecture

The CNNs tested in this project are described, trained and tested using Keras, a high-level neural network API [7], developed as part of the research effort of project ONEIROS (Open-ended Neuro-Electronic Intelligent Robot Operating System) by Franois Chollet, a Google engineer.

2.1   TinyPilotNet

As baseline in our study we use the TinyPilotNet [8] network architecture, developed as a reduction from NVIDIA PilotNet CNN [3] used for self-driving a Car. TinyPilotNet network is composed by a $16 \times 32 \times 1$ pixels image input, followed by two convolutional layers, a dropout layer and a flatten layer. The output of this architecture is formed by two fully connected layers that leads into a couple of neurons, each one of them dedicated to predict steering and throttle values respectively. The input image has a single channel formed by saturation channel from HSV color space (Fig. 2).

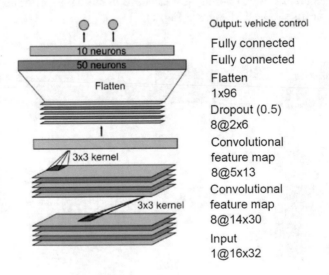

Fig. 2. TinyPilotNet architecture

2.2   Long Short-Term Memory Layers

Long Short-Term Memory (LSTM) layers are included to TinyPilotNet architecture with the aim of improving the CNN driving performance and predict new values influenced by previous ones, and not just from the current input image [9], see Fig. 3.

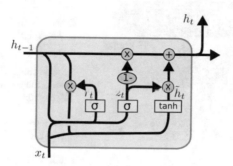

Fig. 3. LSTM cell

These layers are located at the end of the network, previous to the fully-connected layers. During the training, the dataset is used sequentially, not shuffled, in order to allow the network to relate the current output value with the previous ones. This feature makes that current predicted values are influenced by the previous ones, and they do not depend just on the current input image.

2.3   DeepestLSTM-TinyPilotNet

In order to improve the baseline architecture we propose a new network architecture, mainly formed by three 3×3 kernel convolution layers, combined with maxpooling layers, followed by three 5×5 convolutional LSTM layers and two fully connected layers (see Fig. 4).

The LSTM layers produce memory effect, so steering angles and throttle values given by the CNN are influenced by the previous ones.

Convolutional layers extract information from the input image, and pooling layers reduce network to prevent overfitting.

The output is formed by two fully-connected neurons. Each one of them gives the information needed to control the car (steering angle and throttle value).

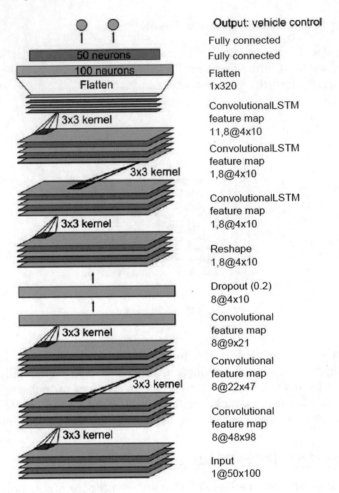

Fig. 4. DeepestLSTM-TinyPilotNet architecture

3   Open-Source Driving Simulator

A dataset with a big amount of images and synchronized output values is needed to train the CNN. Due to the technical complication of doing this in a real environment, an open-source driving simulator is chosen to collect this data. The simulator used to obtain training data is Udacity Driving Simulator [6]. Figure 5 depicts a frame of the simulator.

Some modifications are made in order to collect data from simulator in manual mode. The training data is registered in a Comma-Separated Value file (.csv). Figure 6 shows a fragment of the training dataset used to train the CNN and the types of data registered.

Once the CNN is trained, it is tested driving the car in the simulator, obtaining the new test metrics defined in Sect. 4.2 by collecting driving data. This

Fig. 5. Udacity's driving simulator

center	left	right	steering	throttle	brake	speed
C:\Users\jds96\OneDriv	C:\Users\jds96\OneDriv	C:\Users\jds96\OneDriv	-0.2122216	0.8741225	0	21.06124
C:\Users\jds96\OneDriv	C:\Users\jds96\OneDriv	C:\Users\jds96\OneDriv	-0.2122216	0.8741225	0	21.72626
C:\Users\jds96\OneDriv	C:\Users\jds96\OneDriv	C:\Users\jds96\OneDriv	-0.2122216	0.8741225	0	22.38842
C:\Users\jds96\OneDriv	C:\Users\jds96\OneDriv	C:\Users\jds96\OneDriv	-0.2122216	0.8741225	0	22.83872
C:\Users\jds96\OneDriv	C:\Users\jds96\OneDriv	C:\Users\jds96\OneDriv	-0.2122216	0.8741225	0	23.50869
C:\Users\jds96\OneDriv	C:\Users\jds96\OneDriv	C:\Users\jds96\OneDriv	-0.2122216	0.8741225	0	24.1702
C:\Users\jds96\OneDriv	C:\Users\jds96\OneDriv	C:\Users\jds96\OneDriv	-0.2122216	0.8741225	0	24.82263

Fig. 6. Training data captured from simulator

method allows to analyze the real driving ability instead of a frame-to-frame comparison with human data as ground-truth.

4   Testing Performance

In order to compare the performance of the CNN with other networks, a frame-to-frame comparison is made between CNN steering angle and throttle values and human values as ground truth using only the center camera images.

4.1   Frame-to-Frame Comparison. RMSE Metric

One of the most common test metrics is the Root-Mean Square Error (RMSE) [10] obtained with the difference between CNN steering and throttle predicted values and human given ones. Evaluating the CNN using this method allows to compare the performance with other published networks of the state of the art.

The equation followed is the following:

$$RMSE = \sqrt[2]{\frac{\sum_{t=1}^{T}(\hat{y} - y)^2}{T}}. \tag{1}$$

Driving data collected (y) by a human driver are needed to compare the CNN given values (ŷ) for each frame with the values used by the human driver. This parameter does not evaluate the ability of the network to use previous steering and throttle values to predict the new ones, so the LSTM layer does not make effect and appears underrated in comparison with other CNNs that do not use these kind of layers.

4.2   New Test Metrics

To solve the problem previously mentioned, new quality parameters have been proposed to quantify the performance of the network driving the vehicle. These parameters are measured with the information extracted from the simulator when the network is being tested.

To calculate these parameters, center points of the road -named as waypoints- are needed, separated 2 m as we show in Fig. 7.

Fig. 7. Waypoints of the road in simulator

Center of Road Deviation. One of these parameters is the shifting from the center of the road, or center of road deviation. The lower this parameter is, the better the performance will be, because the car will drive in the center of the road instead of driving in the limits (Fig. 8).

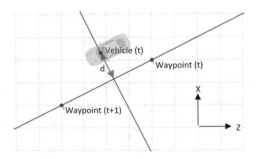

Fig. 8. Distance to the center of the road calculation

To calculate deviation, nearest waypoint is needed to calculate distance between vehicle and segment bounded by that waypoint and previous or next. This waypoint is obtained by using the Eqs. (2) and (3).

Definition 1. *Assume A is the vehicle position and C is Waypoint position. Set:*

$$Waypoint = min\{norm(A, C)\}. \tag{2}$$

Once the nearest waypoint is known, center of road deviation is calculated by using the next equation:

Definition 2. *Assume A is*

$$A = \begin{bmatrix} Waypoint_x(t+1) & Waypoint_z(t+1) & 1 \\ Waypoint_x(t+1) & Waypoint_z(t+1) & 1 \\ Vehicle_x(t) & Vehicle_z(t) & 1 \end{bmatrix}$$

Set:

$$Distance = \frac{abs[norm(A, C)]}{norm(Waypoint(t+1) - Waypoint(t))} \tag{3}$$

Heading Angle. The other parameter is the heading angle. Also, the lower this parameter is, the better the performance will be, because lower heading angle means softer driving, knowing better the direction to follow. To calculate heading angle Eq. (4) is applied (Fig. 9).

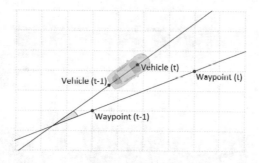

Fig. 9. Heading angle calculation

Definition 3. *Assume veh_vector = Vehicle(t) - Vehicle(t-1), and road_vector = Waypoint(t) - Waypoint(t-1). Set:*

$$heading\ angle = \acos\left(min\left\{1, max\left\{-1, \frac{veh\_vector * road\_vector}{norm(veh\_vector) * norm(road\_vector)}\right\}\right\}\right) * \frac{180}{\pi} \tag{4}$$

5   Experimental Results

In order to determine the performance of the CNNs using the metrics previously established, different experiments are made.

To obtain RMSE parameter, a frame-to-frame comparison between CNN predited values and human given values driving in the simulator is made. RMSE results are displayed in Table 1. The dataset, composed by 17943 frames from center, left and right cameras, used for RMSE calculation is the same that was used for training the CNN.

Table 1. Obtained RMSE metric values.

CNN	RMSE
TinyPilotNet	0.0912475
DeepestLSTM_TinyPilotNet	0.116321

According to Table 1, TinyPilotNet should drive more efficiently than DeepestLSTM_TinyPilotNet due to TinyPilotNet RMSE values are lower than DeepestLSTM_TinyPilotNet ones, so predicted values are more accurate. But this comparison is made frame-to-frame, not feeding the CNN with a driving sequence, so LSTM layers cannot predict values according to the previous ones, so DeepestLSTM-TinyPilotNet hability is underrated in RMSE calculation.

Frame-to-frame comparison is shown graphically in Figs. 10 and 11.

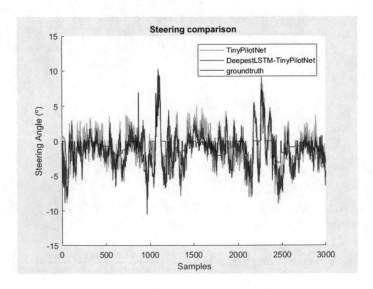

Fig. 10. Steering angles frame-to-frame comparison between DeepestLSTM_TinyPilotNet, TinyPilotNet and human ground-truth

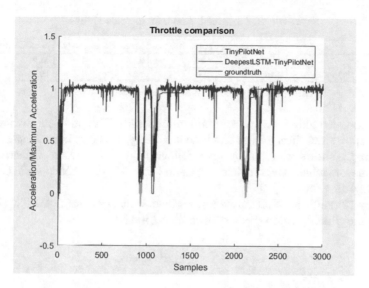

Fig. 11. Throttle frame-to-frame comparison between DeepestLSTM_TinyPilotNet, TinyPilotNet and human ground-truth

RMSE values obtained for the different CNNs can be compared to other networks from the state of the art, taking into account that TinyPilotNet and DeepestLSTM_TinyPilotNet were tested with the same dataset used to train them.

For the CNNs proposed in this paper RMSE results are very close to other state of the art networks such as PilotNet [3], AlexNet [11] and VGG-16 [12], as can be seen in Table 2.

Table 2. RMSE obtained values compared with state of the art CNNs.

CNN	RMSE
TinyPilotNet [8]	0.0912475
DeepestLSTM_TinyPilotNet	0.116321
AlexNet [11]	0.1299
PilotNet [3]	0.1604
VGG-16 [12]	0.0948

New test metrics values are obtained from driving data extracted from the simulator when the car is controlled by the CNN, using the Eqs. (2), (3) and (4). Obtained metric values are displayed in Table 2.

As can be seen in Table 3, DeepestLSTM_TinyPilotNet is able to reduce the mean and maximum distance to the center of the road, which leads to safer

Table 3. Obtained new test metrics values.

CNN	Mean distance	Max. distance	Mean heading angle	Max. heading angle
TinyPilotNet	1.07	7.17	3.38	44.91
DeepestLSTM_TinyPilotNet	0.72	5.00	2.15	26.35

driving, and also reduces mean and maximum heading angle values, which means the car follows the direction of the lane smoothly, without big changes.

Figure 12 shows the trajectories followed by the car in Udacity's driving simulator when the vehicle is leaded by TinyPilotNet and DeepestLSTM_TinyPilotNet

Figure 12 reaffirms what Table 3 quantifies, that DeepestLSTM_TinyPilotNet drives more directly and centered than TinyPilotNet.

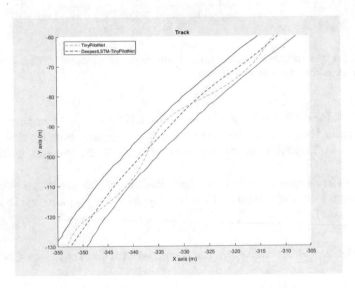

Fig. 12. Fragment of track showing TinyPilotNet and DeepestLSTM_TinyPilotNet car trajectories.

6   Conclusions

This paper describes a new testing metrics system that allows for a better knowledge of the CNN driving ability because it tests the network by doing the real task it was trained for.

The inclusion of Long-Short Term Memory (LSTM) layers in the network architecture produce a smoother driving by taking into account the previous steering and throttle values given by the CNN (a sequence), and not just a single moment. DeepestLSTM_TinyPilotNet drives nearest to the center of the road and straighter than the original TinyPilotNet, as can be seen in Fig. 12.

Acknowledgment. This work has been partially funded by the Spanish MINECO/ FEDER through the SmartElderlyCar project (TRA2015-70501-C2-1-R), the DGT through the SERMON project (SPIP2017-02305), and from the RoboCity2030-III-CM project (Robótica aplicada a la mejora de la calidad de vida de los ciudadanos, fase III; S2013/MIT-2748), funded by Programas de actividades I+D (CAM) and cofunded by EU Structural Funds.

References

1. Society of Automotive Engineers self-driving vehicles classification. https://www. cnet.com/roadshow/news/self-driving-car-guide-autonomous-explanation/
2. Vehicle detection using machine learning and computer vision. https:// towardsdatascience.com/teaching-cars-to-see-vehicle-detection-using-machine-learning-and-computer-vision-54628888079a
3. Bojarski, M., et al.: End to end learning for self-driving cars. NVIDIA (2017). arxiv.org/pdf/1604.07316.pdf
4. Europilot. Marsauto. https://github.com/marsauto/europilot
5. Gazebo. http://gazebosim.org/blog/car_sim
6. Udacity self-driving car simulator project. https://github.com/udacity/self-driving-car-sim
7. Keras documentation. https://keras.io/
8. Shao, Y.: End-to-end learning for self-driving cars (2017). https://github.com/ ymshao/End-to-End-Learning-for-Self-Driving-Cars
9. Convolutional, Long Short-Term Memory, fully connected deep neural networks. https://static.googleusercontent.com/media/research.google.com/es// pubs/archive/43455.pdf
10. Chi, L., Mu, Y.: Deep steering: learning end-to-end driving model from spatial and temporal visual cues (2017). https://arxiv.org/pdf/1708.03798.pdf
11. Krizhevsky, A., Sutskever, I., Hinton, G.E.: Imagenet classification with deep convolutional neural networks. In: Advances in Neural Information Processing Systems, pp. 1097–1105 (2012)
12. Simonyan, K., Zisserman, A.: Very deep convolutional networks for large-scale image recognition, arXiv:1409.1556 [cs], September 2014

Study of Obstacle Avoidance Strategies for Efficient Autonomous Navigation of Wheeled Robots in Unknown Environments

Julián Cristiano[✉], Hatem A. Rashwan, and Domènec Puig

Intelligent Robotics and Computer Vision Group,
Department of Computer Science and Mathematics,
Rovira i Virgili University (URV), Tarragona, Spain
julianefren.cristiano@urv.cat
http://deim.urv.cat/%7Erivi/

Abstract. This paper studied several obstacle avoidance strategies proposed within recent state of the art for autonomous navigation of wheeled robots to efficiently deal with unknown environment. Some initial experiments have been performed in this paper to identify the most important features, (e.g., obstacle representation, efficient control of robot's direction, etc.) that an obstacle avoidance strategy must include in order to guarantee an effective autonomous navigation system for wheeled robots in unknown environments.

Keywords: Mobile robots · Obstacle avoidance · Wheeled robots

1 Introduction

The autonomous navigation of wheeled robots in unknown and dynamically changing environments is a complex task which, requires a robust obstacle avoidance strategy to be able of detecting moving obstacles while ensuring an efficient navigation in terms of energy consumption along with fast execution time [1–8]. The robots perceive the environment from the data provided by their onboard sensors [8]. There is a wide variety of sensors used to detect the obstacles: light detection and ranging (LIDAR) sensors, radar sensors, ultrasonic sensors, image based systems utilizing monocular, stereo cameras and RGB-D cameras.

Wheeled robots are commonly used for explorations tasks. Several robotics platforms have been built in the last years. These wheeled robots are available commercially with different sizes and different technical specifications, to name few, computational processing capability, onboard sensors available, etc.

The hardware available and power consumption requirements are important aspects to consider in the design of navigation systems as them determine the final robustness of the system. For instance, for the navigation systems of autonomous cars, a robust system is required in order to assure the safety of

© Springer Nature Switzerland AG 2019
R. Fuentetaja Pizán et al. (Eds.): WAF 2018, AISC 855, pp. 44–55, 2019.
https://doi.org/10.1007/978-3-319-99885-5_4

the people inside and outside the car. Therefore, reliable and redundant sensors must be used.

Consequently, this paper attempts to study several obstacle avoidance strategies for autonomous navigation of wheeled robots to efficiently deal with unknown environment. Some initial experiments have been performed to identify the most important features that an obstacle avoidance strategy must include.

This paper is organized as follows: Sect. 2 summarizes some relevant recent works. Section 3 presents some important aspects that an efficient obstacle avoidance strategy must include. Some experimental results are presented in Sect. 4 and their respective discussion in Sect. 5. Finally, conclusions and future work are given in Sect. 6.

2   Obstacle Avoidance Strategies

The obstacle avoidance strategies must include the ability to properly detect, characterize and avoid the obstacles surrounding the robot in order to allow the navigation system reaches its final destination. The ability to perform robust obstacle avoidance in unknown environments using sensory information that assures a robust navigation is a challenging task.

Most of the systems proposed in the literature provide solutions for dealing with static objects, however, the big challenge appears whenever there are objects moving within the workspace and also in presence of uneven terrain.

The robot perceives the objects through its sensory information, therefore according to the accuracy of the sensors used by the robot a higher or lower accuracy in the geometric representation of the environment can be obtained. The geometric information, such as the shape of the obstacle and the robot, the robot's kinematic information [1], etc., allows to identify the gap between the robot and the obstacle in order to compute an optimal collision free trajectory for an efficient navigation.

In order to achieve robust navigation, a system able to detect and avoid quickly the obstacles is required. Thus, it is required to gather useful data using sensors with a good resolution and responsiveness and also to have a system with an adequate computational capacity to analyse the data [2].

Several systems have proposed using diverse types of wheeled robots and sensors. However, there is a trade off between the available sensory information, the robot's computational capacity and the desired accuracy according to the requirements of the obstacle avoidance system. In this paper we present the more used methods and their advantages and limitations.

Most of the works found within the state of the art were proposed for wheeled robots navigating on flat terrain, however, one of the big challenges is the navigation on uneven terrain, including small obstacles or sloped terrain.

The perception of the robot can be limited due to the sensors used. Therefore, it is very important to take in consideration the physical constraints of the perception system in order to overcome some possible unexpected situations.

The more relevant obstacle avoidance strategies can be divided into three categories: model-based methods, fuzzy logic methods and reactive methods based on machine learning.

2.1 Model-Based Methods

The methods belonging to this category use all the available information provided by the sensors. The complete kinematic model of the robots is used in order to define a precise mathematical representation of the robot and its surrounding area [1]. These methods require to accurately know the robot's kinematic model, an accurate representation of the obstacle around the robot, etc.

In [2], the authors propose a system for navigation of a wheeled robot on sloped terrain. Their system differentiate stairs from sloped terrain enabling the navigation of wheel robot on sloped terrain. 3D mapping of the environment was used in order to detect and characterize it, allowing the robot to perform an efficient navigation.

These methods guarantee an accurate obstacle avoidance strategy, which also favours the optimal navigation in unknown environments. However, good hardware specifications are required to deal with the computational complexity of the system. For instance, the obstacle being recognized is limited by the on-board sensors used to percept the robot's surrounding area.

2.2 Fuzzy Logic Methods

These methods take advantage of the fuzzy logic theory in order to develop reactive behaviours that help the robot to overcome some obstacles [3,4]. The implementation of these methods is simple in comparison with the model based methods and the obtained reaction time is small which guarantees a fast system response. However, the main disadvantage of these methods is that can not navigate efficiently in complex and unknown environments. It is due to the fact that some of the parameters must be defined manually and hence the system can not adapt itself to new unexpected situations.

The fuzzy logic methods work properly for static obstacles. However, the behaviour of these systems in complex environments require to put several constraints in the system in order to guarantee the appropriate interaction with the obstacles.

2.3 Reactive Methods Based on Machine Learning

These methods use machine learning algorithms to learn the reactive behaviour that must be produced by the robot in order to efficiently overcome with the obstacle [5–7]. Initial data demonstrating the desired behaviour is required by the system in order to train it. The data must be provided by an expert, for instance through demonstration by using tele-operation of the robot or by pre-defining some rules in a similar way to the fuzzy logic methods. The learning process can be time consuming, specially if the system considers extensive sensory information.

The main disadvantage of these methods is that in order to overcome any obstacle, a big training dataset with all the possible configurations is required, which is not viable from the practical point of view. If the system experiments some cases that were not learned in advance, the system can behave in an unpredictable way.

3   Efficient Obstacle Avoidance Strategy

Based on the aforementioned works, this section presents some of the most important aspects that must contain an efficient obstacle avoidance strategy in order to overcome with different obstacles around the robot while it navigates in unknown terrain.

The data provided by the perception system from the environment sensing is the most important element to detect and characterize the obstacle's shape and hence to effectively compute a collision free trajectory. Therefore, it is crucial to make an efficient representation of the obstacles and the ground. This information can be used to define multilayer maps as decribed in [2]. These maps are an useful representation to characterize the surrounding area of the robot. It is an effective way to represent the complete scenario around the robot and also can be used for an efficient generation of collision free trajectories and hence autonomous navigation, for instance by using a model based method. The maps accurately allow to compute the minimum gap distance between the robot and any obstacle around it in different direction according to the used sensors.

It has been identified from the literature review that data gathered by a 3D lidar is the best option. This sensor works both in indoor and outdoor environments and it can detect and allow the accurate shape characterization of obstacles located to near or far distances from the robot. In addition, this sensor can be used to detect accurately uneven terrain surfaces.

To simplify the obstacle representation, it is important to use different resolutions for the obstacles representation and also according to limitation imposed by the hardware resources available. Thus, for instance octomap can be used to perform the 3D modelling of the environment around the robot [9]. Octomap is an integrated framework based on octrees for the representation of three-dimensional environments.

Initial experiments presented in the next section indicate that a robust obstacle avoidance strategy can be develop using a robust perception system. This perception system must be used to characterize the obstacles and also the ground.

4   Experimental Results

4.1   Simulation Environment

This section presents some initial experiments performed using a simulated environment to test different sensors used commonly for autonomous navigation.

Fig. 1. Pioneer 3-AT with a 3D lidar Velodyne HDL-64E

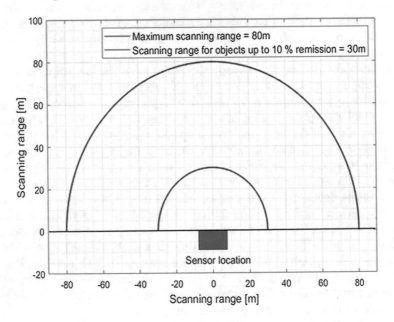

Fig. 2. Scanning range for the 2D lidar LMS291 [10].

The goal was to assess the performance of the proposed system. Different testing scenarios were designed in order to study obstacle avoidance strategies with different sensors. Simulation experiments have been conducted with Webots [11]. A robust simulator based on the Open Dynamics Engine library[1].

The robot model used to perform the experiments was the Pioneer 3-AT [12], a small four-wheel robot with a weight of 12 kg (see Fig. 1). The robot was developed by Omron Adept MobileRobots.

We have decided to compare the response of the obstacle avoidance system with three different sensory input. Thus, three sensors were studied to detect the obstacles in the robot's surrounding area. These sensors are: a 3D lidar sensor, a 2D lidar sensor and a range camera. These sensors are used commonly for

---

[1] http://www.ode.org/.

autonomous navigation. The 2D lidar used for the experiments was the SICK LMS291. The scanning range of the 2D Lidar sensor is shown in Fig. 2. The 3D lidar used in the experiments was the Velodyne LIDAR HDL-64E[2]. It has a range of 120 m and supports 64 channels with a 360° horizontal field of view as well as a 26.9° vertical field of view. The range camera used was the Kinect camera v1. The pioneer robot was equipped with the three sensors in the simulated environment in order to study the data provided by each sensor with the aim of achieving efficient obstacle avoidance.

The data gathered by the Kinect and the 3D lidar sensor can be used to create a 3D map of the robot's surrounding area. Later, this information is divided in horizontal lines that represent different heights also called layers. The idea is to detect the objects in different layers. Multilayer maps, as described in [2], were computed to characterize the robot's surrounding region. These maps are an effective way to represent the complete scenario around the robot and also can be used for the efficient generation of collision free trajectories and hence autonomous navigation. The data generated by the 2D Lidar just provide information to a fixed height. Thus, this sensor can not be used for robust navigation in unknown terrain.

Most of the wheeled robot's use odometry to estimate their current position. However, in presence of uneven terrain, the odometry is not accurate, therefore it is highly recommended to use a Global Positioning System (GPS) sensor in outdoor environments or Bluetooth signals for indoor environments [13].

4.2   Simulated Scenarios

Two simulated scenarios have been implemented in order to test the most important functionalities that an efficient obstacle avoidance strategy must incorporate in order to efficiently overcome with the obstacles and the rough or uneven terrain. The two scenarios implemented are described below. A video showing some experimental results in the simulated environment using the three studied sensors can be found on the companion website[3].

Testing Scenario 1: The first testing scenario includes four walls and one obstacle. The obstacle is located around the surrounding area of the robot and the sensory information provided by different sensors is used to study how the robot perceives the obstacles with each sensor. In this scenario the navigation ground was flat. Figures 3 and 4 show the experimental results using the 2D Lidar sensor. Figures 5 and 6 show the experimental results using the 3D Lidar sensor.

Testing Scenario 2: This scenario includes the same obstacle included in scenario 1 and also some obstacles on the floor to represent a uneven terrain. The

---

[2] http://www.velodynelidar.com/hdl-64e.html.
[3] https://youtu.be/kRPXGi47m-o.

Fig. 3. This figure shows the pioneer robot moving in the simulated environment using a 2D Lidar LMS291.

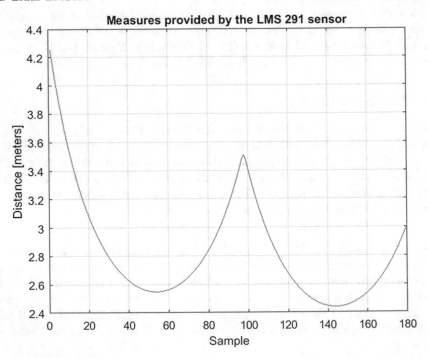

Fig. 4. This plot represents the data gathered by the robot shown in Fig. 3. It can be appreciated that the robot does not detect the obstacle in from of it as the obstacle's height is lower than the height of the Lidar sensing area.

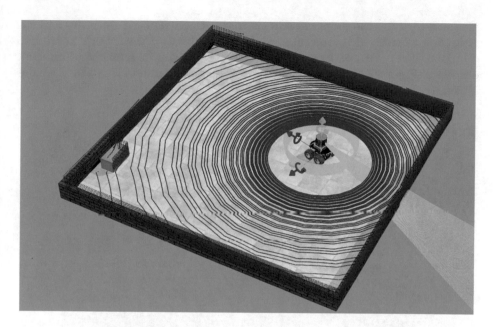

Fig. 5. This figure shows the pioneer robot moving in the simulated environment using a 3D Lidar Velodyne HDL-64E and the projection of each lidar's sensing layer.

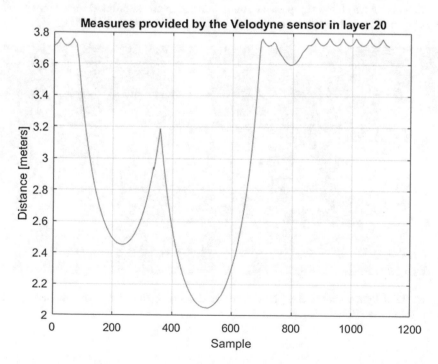

Fig. 6. This plot represents the data gathered by the robot shown in Fig. 5.

idea was to study the proposed scheme to determine how the obstacles and the ground are perceived by the robot and how them affect the robot's navigation direction. The goal was to study the system response in presence of different obstacles. Figures 7, 8 and 9 show the experimental results using the 3D Lidar sensor.

Fig. 7. This figure shows the pioneer robot moving in the simulated environment with diverse obstacles and rough terrain using a 3D Lidar Velodyne HDL-64E.

Fig. 8. This figure shows the pioneer robot detecting the rough terrain using a 3D Lidar Velodyne HDL-64E

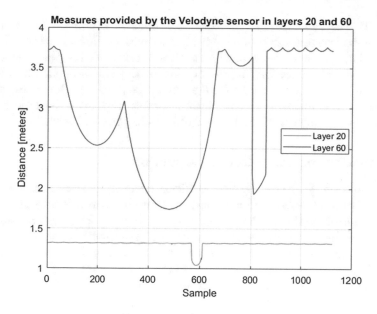

Fig. 9. This plot represents the data gathered by the robot shown in Figs. 7 and 8. The data represent two layers, layer 20 is used to detect the ground shape and layer 60 is used to detect the obstacles located in the surrounding area of the robot.

5 Discussion

Some important requirements for an efficient obstacle avoidance system have been identified from the literature review and from some initial experiments performed. These are summarized and organized in two categories presented below.

5.1 Obstacle and Environment Representation

Artificial perception for autonomous navigation is an important process by which an intelligent system translates sensory data for two important tasks ground detection and obstacle detection. The ground detection aims to know in advance the floor features, while the obstacle detection wants to detect and characterize the static and/or moving obstacles in the surrounding area of the robot. An accurate representation of the obstacle is an important component of autonomous navigation systems [8], as it serves as input to the generation of collision free trajectories. However, it can be computationally expensive and therefore it is required to define a system that can increase or decrease the obstacle and ground model according to the required accuracy.

The environment representation is also important for autonomous navigation, for instance multilayer maps and traversable maps as described in [2] are an useful source of information for the autonomous collision free navigation system and also in order to facilitate motion planning tasks.

Full 3D sensory information is required to achieve effective autonomous navigation able to overcome unknown environments with static and moving obstacles. The 3D lidar scanners are the most suitable sensors, since they provide accurate and quick reliable data. The representation of the environment and obstacles must be determined according to the accuracy required by the obstacle avoidance system. For instance, the resolution used to represent the object can be lower for distant objects while for near objects it must be higher.

5.2   Efficient Control of Robot's Direction

The information generated by the artificial perception system to detect and characterize the obstacles and environment must be used to efficiently compute the best motion that the robot requires to navigate. The system must take advantage of the hardware resources available in the robot and also its kinematic structure. Thus, the minimum gap between the robot and the obstacle and the ground shape must be used to generate the motor command to the robot motors. For instance, using a model based method.

6   Conclusions and Future Work

This paper presents a study of diverse obstacle avoidance strategies for efficient autonomous navigation of wheeled robots in unknown environments. Some initial experiments using a 2D lidar, a 3D lidar and a Kinect camera have been performed in simulation in order to identify the most important features that an effective system must include. The obstacle avoidance strategy must be selected from the functionalities required by the system and also according to the hardware available. In order to navigate in unknown environments with moving objects and rough terrain, it is needed that the perception system for object and environment detection and characterization represents efficiently the robot's surrounding area in order to allow it a collision free navigation and also to facilitate the motion planning tasks. In order to assure an efficient navigation in unknown terrain, a 3D lidar sensor provides the more reliable information in order to understand the environment around the robot through multilayer representation.

Future work will include the implementation of the system in a real pioneer robot in order to test the performance of the system in a real scenario. Also, to continue studying the best representation of the data generated by the 3D Lidar in order to facilitate its analysis and use in indoor and outdoor navigation.

References

1. Mujahed, M., Fischer, D., Mertsching, B.: Admissible gap navigation: a new collision avoidance approach. Robot. Auton. Syst. **103**, 93–110 (2018). https://doi.org/10.1016/j.robot.2018.02.008
2. Wang, C., Meng, L., She, S., Mitchell, I.M., Li, T., Tung, F., Wan, W., Meng, M.Q.H., de Silva, C.W.: Autonomous mobile robot navigation in uneven and unstructured indoor environments. In: 2017 IEEE/RSJ International Conference on Intelligent Robots and Systems (IROS), pp. 109–116 (2017). https://doi.org/10.1109/IROS.2017.8202145
3. Minguez, J., Montano, L.: Extending collision avoidance methods to consider the vehicle shape, kinematics, and dynamics of a mobile robot. IEEE Trans. Robot. **25**(2), 367–381 (2009). https://doi.org/10.1109/TRO.2009.2011526
4. Pothal, J.K., Parhi, D.R.: Navigation of multiple mobile robots in a highly clutter terrains using adaptive neuro-fuzzy inference system. Robot. Auton. Syst. **72**, 48–58 (2015). https://doi.org/10.1016/j.robot.2015.04.007
5. Antich Tobaruela, J., Ortiz, R.A.: Reactive navigation in extremely dense and highly intricate environments. PLoS ONE **12**(12), e0189008 (2017). https://doi.org/10.1371/journal.pone.0189008
6. Duguleana, M., Mogan, G.: Neural networks based reinforcement learning for mobile robots obstacle avoidance. Expert. Syst. Appl. **62**, 104–115 (2016). https://doi.org/10.1016/j.eswa.2016.06.021
7. Xia, C., El Kamel, A.: Neural inverse reinforcement learning in autonomous navigation. Robot. Auton. Syst. **84**, 1–14 (2016). https://doi.org/10.1016/j.robot.2016.06.003
8. Asvadi, A., Premebida, C., Peixoto, P., Nunes, U.: 3D Lidar-based static and moving obstacle detection in driving environments: an approach based on voxels and multi-region ground planes. Robot. Auton. Syst. **83**, 299–311 (2016). https://doi.org/10.1016/j.robot.2016.06.007
9. Hornung, A., Wurm, K.M., Bennewitz, M., Stachniss, C., Burgard, W.: OctoMap: an efficient probabilistic 3D mapping framework based on octrees. Auton. Robot. **34**, 189–206 (2013). https://doi.org/10.1007/s10514-012-9321-0
10. LMS291 datasheet. https://cdn.sick.com/media/pdf/9/49/849/dataSheet_LMS291-S05_1018028_es.pdf
11. Michel, O.: Cyberbotics Ltd. Webots^{TM}: professional mobile robot simulation. Int. J. Adv. Rob. Syst. **1**, 39–42 (2004)
12. Pioneer 3-AT datasheet. http://www.mobilerobots.com/Libraries/Downloads/Pioneer3AT-P3AT-RevA.sflb.ashx
13. Schwiegelshohn, F., Wehner, P., Werner, F., Gohringer, D., Hubner, M.: Enabling indoor object localization through Bluetooth beacons on the RADIO robot platform. In: 2016 International Conference on Embedded Computer Systems: Architectures, Modeling and Simulation (SAMOS), Agios Konstantinos, pp. 328–333 (2016). https://doi.org/10.1109/SAMOS.2016.7818366

Combination of Semantic Localization and Conversational Skills for Assistive Robots

Daniel González-Medina$^{(\boxtimes)}$, Cristina Romero-González$^{(\boxtimes)}$, and Ismael García-Varea$^{(\boxtimes)}$

University of Castilla-La Mancha, Campus Univ. s/n., 02071 Albacete, Spain
{Daniel.Gonzalez,Cristina.RGonzalez,Ismael.Garcia}@uclm.es

Abstract. The recognition of objects and their features is a fundamental task for social robots that could be improved with the combination of different sources of information, such as the ones provided by visual or speech understanding systems. In this paper, we present a first approach to fusion semantic localization and conversational skills for social robots which may act as assistants. Our solution is based on a mobile robot that is able to detect and recognize objects from an environment and store them in its base of knowledge to later act as an assistant for any user who is searching for any object. In the conversation the robot tries to help the user to find a specific object depending of the location and the features of the object which is looking for. The proposal has been empirically evaluated within a research lab where the robot recognizes objects in the environment and the users require, by means of speech commands, finding suitable objects that are placed in the environment.

Keywords: Human-robot interaction · Assistive robotics
Smart homes · Artificial vision · Speech recognition

1 Introduction

Human-Robot Interaction (HRI) has become one of the most considerable research fields with a large range of real-world applications in the academic community, in labs, in technology companies and even the media [6]. Thanks to the current technologies, the robots can engage in conversation with humans while automatically perceiving the emotional state of users, recognizing their gestures, or even understanding their body language [13].

At the moment, one of the key themes and one of the biggest challenges in the HRI field that the society faces today is the provision of assitive robots in smart homes [3] for both elder people or people with disabilities, and any person in their day-life activities. In such setting, assistive robotics provide early detection of problems and emergencies, and are a crucial interface to attempt the immediate specific needs of the person living in such homes. Such assistances have to be defined in the long-term and, to support that, the robotic systems

© Springer Nature Switzerland AG 2019
R. Fuentetaja Pizán et al. (Eds.): WAF 2018, AISC 855, pp. 56–69, 2019.
https://doi.org/10.1007/978-3-319-99885-5_5

must be capable of acquiring new concepts and adapt to new situations to solve tasks by exploiting its past experiences. Based on this premise, the LifeBots (Lifelong Technologies for Social Robots in Smart Homes) research project aims at developing a lifelong system that facilitates and improves the integration of robotics platforms in smart homes to support elder people and people with disabilities [1]. Among other tasks, the robot must be able to (1) identify and recognize the key elements in its environment, (2) respond to different commands and (3) understand natural language. In this paper, we propose a base model that solves these three tasks where an assistive robot can localize objects and interact with people in a home-like environment.

More specifically, our proposal should result in a robotic system that automatically identifies and recognizes objects from an environment, its locations and its features, and stores them in its persistent base of knowledge, with the purpose of being able to answer questions about specific objects of the environment. The robot has to be able to perform visual scene understanding as well as speech recognition and synthesis.

The system has been evaluated with an use case that involves these three tasks, where a robot behaves as an automatic assistant robot to localize objects from an known indoor place. In Fig. 1 the overall scheme for the use case that has been defined to evaluate our proposal is presented. This use case has been evaluated in a real-world scenario and involving different people asking the robot for various objects. The robot has been able to interact with the users with a natural conversation, meeting their requirements, and answering the questions and doubts of the users.

The remainder of the article is organized as follows. A brief overview of the LifeBots research project is given in Sect. 2. Then, Sect. 3 describes the developed system, where we define all the modules employed and their role in the solution of the problem. A detailed explanation of the use case to evaluate the system is included in Sect. 4. Finally, the results achieved and the obtained conclusions are shown in Sects. 5 and 6, respectively.

2   LifeBots Research Project

Nowadays, it is clear that the inclusion of robotics in everyday life will have a large impact and it will probably be totally established in few decades [12]. One of the targets, while robotics develops, is that domestic day-life activities related to personal assistance will be provided by robotic platforms, incorporating them into indoor scenarios such as smart homes. In this context, assistive platforms are required to reduce or eliminate the needs of older people living alone or individuals with physical limitations and disabilities to continue living independently in their own homes. Furthermore, it has been demonstrated that assistive robotics is also a crucial interface to the person living in the smart home [7].

The main goal of LifeBots research project is focused on covering this area of the social robotic interaction. Specifically, it relies on long-term assistance where the robotic system should acquire new concepts and solve tasks by exploiting its

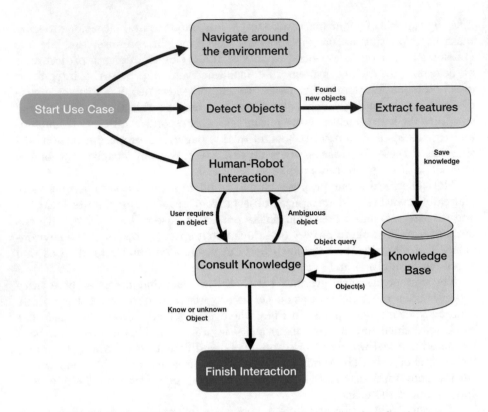

Fig. 1. Overall diagram of the use case for the proposal evaluation.

past experience. Thus, the robot does not have to solve any task from scratch but supported from previous experience. The LifeBots project includes multiple tasks to develop, and different use cases to carry out and evaluate those tasks.

One of the tasks is the combination of semantic location with conversational skills to merge different sources of information with the purpose of improving the recognition of objects, places or speech based interactions. The system that we present in this article is a prototype model for this task.

3   System Proposal and Description

One of the challenges for social robots is interacting properly with humans. In order to do it, the robots need to be able to maintain a conversation with people holding high-level dialogues, and to this end, the robot must be capable of recognizing and interpreting human speech, including affective speech, discrete commands and natural language [10].

The main goal of our proposal is to provide a robotic system with the capability to automatically recognize objects and its features at any kind of indoor environment and store them in its knowledge base. This information will be useful later to act as an assistant to answer any question about the location of an object based on its features. The selected features for the objects are shown in Table 1, and the type of questions that the robot could be capable of answering are questions about where a specific object is placed, it could also respond to commands about bringing some objects. Depending on the object and the information about the object that the user is asking for, the robot will consult its knowledge base and will answer accordingly.

Table 1. Features of the objects.

Feature	Description
Type	Object type, e.g.: cup, sofa, laptop, etc.
Colour	Main colour of the object, e.g.: blue, red, etc.
Location	Location from the semantic map in the Fig. 5
Time	The moment when the object was detected

The architecture of the developed system to carry out our proposal is described in Fig. 2, where we show a diagram detailing the modules and submodules of the system, and how they are connected to each other. Regarding the software architecture, we have relied on the an open-source framework designed for robotics programming. With this platform, we control all the communications between the modules and we manage the interaction between them.

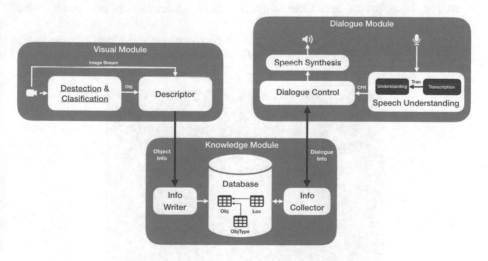

Fig. 2. Graphical description of the proposal.

The system is composed of three main modules, which are described in the following subsections.

3.1 Visual Module

In order to detect the objects that are placed in the environment and to classify them, this module uses a camera to capture what the robot is looking at and transforms what it is watching in the scene in valuable information. Besides the object detection and classification, this module also extracts the features that describe that object (see Table 1). This module is divided into two sub-modules:

1. **Detection and classification:** this module deals with the detection of potential objects in the environment using a image stream from a camera as a source of information. To this end, it detects candidate objects and classifies them using a deep learning pretrained model. Specifically we have used the well-known YOLO system [11]. An example of object detection is presented in Fig. 3, where we can observe how the objects in the scene are successfully detected and classified. The output of this module is the class of the detected objects and their corresponding bounding boxes.

2. **Description:** as a result of the object detection and classification, this module takes the object and the image from the camera where the object was detected, and extracts its features. The features extracted are the location where the object is placed in the environment and the dominant colour of the object. To get the location we use the position of the object and the semantic map of the environment, and to get the dominant colour we use a K-means algorithm [8] over the center of the image. We calculate five clusters based on pixel RGB-colours and we take the centroid of the cluster with the biggest amount of pixels as the dominant colour. We can see an example in the Fig. 4. We also convert the resulting dominant RGB-colour into a colour name such as red, blue, green, etc. The dominant colour extraction is performed using the *OpenCV* library [2].

After taking the location and the colour of the object, the module sends the information to the knowledge module to store it.

Fig. 3. Object detection and classification.

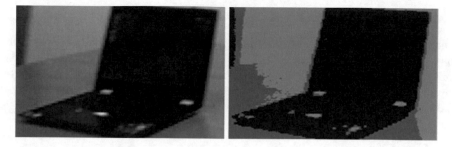

Fig. 4. *K*-means of 3 clusters applied to the pixel RGB-colours of an example image, where black is the dominant colour.

3.2   Dialogue Module

To manage the dialogue to naturally interact with the users, this module has been provided with speech recognition and understanding capabilities to maintain a conversation with a human. This module is divided into tree sub-modules:

1. **Speech understanding:** in other to understand what the user is talking, this module manages the input audio when a person speaks. It takes the audio with a microphone as an input and it extracts the CFR (Command Frame Representation) [9] as an output. This task is performed in two separate stages. First, an automatic, continuous speech recognizer is employed to obtain the speech transcription from the user utterance. For this task, we have relied on the *Google Speech Recognition* system [14] that can accurately transcribe speech in a real environment. Once the transcription is available, the next stage deals with the interpretation of the words uttered by the user. Here, to obtain a semantic representation of the transcribed sentences (CFR), it uses a parsing of the text transcription using the *SENNA* [4] toolkit in order to obtain elements like part of speech, name entities or the semantic role labelling (SRL) of the sentence. The output of *SENNA* is processed following a rule-based approach to obtain a CFR. Finally, the CFR is sent to the dialogue control module.
2. **Dialogue control:** as a result of the speech understanding module, a suitable representation of the user requirement is obtained to look up the robot knowledge base with the aim of formulating a proper answer to say to the user. The answer is constructed depending on the knowledge of the robot and the requirements of the user. When there is a mistake in the CFR because the speech understanding module did not understand the user, an answer is sent to the speech synthesis module asking the user to repeat what he said. The module can take into account a maximum of three mistakes before finishing the interaction.
3. **Speech synthesis:** in this module the answer to the user is said through the speakers, using the *Google Text-to-Speech* [5] library, which transforms text to speech.

3.3   Knowledge Module

The main component of this module is a *MySQL* database where all the information coming from the visual and audio modules is stored as knowledge of the robot, such as the objects, their features and their location with regard to the environment. This module has two sub-modules that transform the input information into *MySQL* queries to consult or to write into the database, respectively.

4   Use Case

The use case that we have developed to evaluate our system proposal is based on a robot that previously has been able to navigate around an indoor environment. This environment is a laboratory scenario (turned into a home-like flat), which is composed of three rooms, and its dimensions are 5.95×5.65 m, the overall size of the working scenario is over 33 m^2. The three possible locations correspond to semantic places in this home-like scenario. These locations are: a *dinning room* (big table and chairs), a *kitchen* (sink and coffee machine), and a *living room* (couch, coffee table, and bookshelf). Some pictures of the scenario and its setup are shown in Fig. 5.

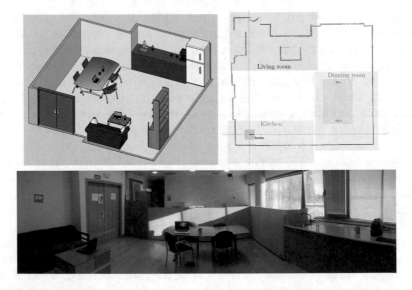

Fig. 5. A 3D model of the environment, its semantic map, and a panoramic picture.

The robot employed is a differential-drive PeopleBot robotic platform fitted with additional sensors and computing capabilities (see Fig. 6). More specifically, the robot has been equipped with an RGB-D camera and a directional microphone that improves the speech recognition performance. The initial PeopleBot

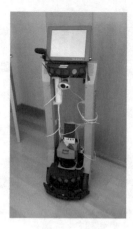

Fig. 6. Robotic platform used in the use case.

platform includes the navigation Hardware package, whose main device is a SICK LMS200 2D laser. It has also been upgraded with an Intel NUC processor.

While the robot is navigating around the environment it is automatically completing and increasing its base of knowledge with the objects and their features. Later, when a user comes into the scenario and needs assistance to find or bring an object, they might ask the robot.

The conversation that the robot is able to hold consists of questions made by the person to the robot about the localization of objects. Depending on the knowledge of the robot it will respond with different answers. Three different scenarios were tested:

1. If the robot knows the object → it will answer with the location.
2. If the robot doubts between different objects → it will answer with a question asking about what object the person is looking for based on its knowledge.
3. If the robot does not know the object → it will answer it does not know that object.

If the robot doubts about the object the person is asking for, it will use the different features of the objects to find out the object the person is asking for. The use case finishes when the robot recognizes the object which the person is asking for or the robot does not know the object.

In Fig. 7 we can observe a graph with the pipeline of the dialogue that the robot and the person could hold, where the robot recognizes the user requirement, consults its knowledge base and responds with an appropriate answer to the user.

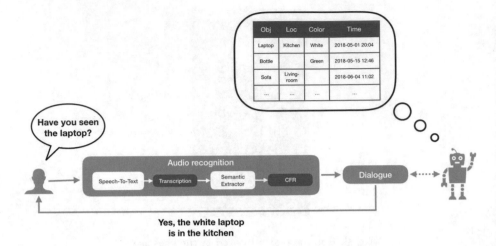

Fig. 7. Pipeline of the dialogue between the robot and an user.

5 Experiments and Results

The experimentation we have carried out to test the proposal was performed during several days and considering different objects in the scenario and various interactions between the robot and multiple users. Specifically, the experiments consisted of: first, the robot navigated around the environment and stored in its knowledge base the objects that it found; then, different users interacted with the robot, where each user could ask the robot about the objects in the environment. During the human-robot interaction process, the robot has to determine which object the user is asking for and try to help him based on its base of knowledge.

Despite that the experiment involved the complete development of the use case, we focused the experimentation on the evaluation of the specific tasks or capabilities. Concretely, we validated the detection and classification accuracy of the objects from the visual module, as well as its performance on the dialogue with the users. This last part was evaluated by means of a satisfaction test that the different users fulfilled after finishing the interaction experience.

5.1 Object Detection and Classification Results

Here, we evaluate the visual module which involved the detection, classification and colour extraction from the objects. We calculate the precision and the recall of these tasks which determines the base knowledge of the robot to later answer the question of the user.

The objects present in the environment and their features (location and colour) are shown in Table 2. In Table 3, the objects detected by the robot are shown with the classification and the extracted colour.

The visual classification results are shown in Table 4, where precision and recall measures are presented for object classification and location assignation

Table 2. Objects present in the environment.

Object type	Location	Colour
Cup	Dinning-room	Blue
Cup	Dinning-room	Green
Bottle	Dinning-room	Transparent
Laptop	Dinning-room	Black
Diningtable	Dinning-room	Brown
4 Chairs	Dinning-room	Blue
Laptop	Living-room	Gray
Cup	Living-room	Orange
Bottle	Living-room	Transparent
Sofa	Living-room	Blue
2 Books	Living-room	Gray

Table 3. Objects detected by the robot.

Object type	Location	Colour
Cup	Dinning-room	Blue
Cup	Dinning-room	Gray
Bottle	Dinning-room	Gray
Laptop	Dinning-room	Black
Diningtable	Dinning-room	Green
2 Chairs	Dinning-room	Blue
Chair	Dinning-room	Black
Bowl	Dinning-room	Black
Laptop	Living-room	Gray
Cup	Living-room	Gray
2 Bottles	Living-room	Gray
Sofa	Living-room	Blue
2 Books	Living-room	Gray

to each object, and the accuracy of the colour assigned to each object. As we can observe, both classification and location assignation are accurate, which is necessary since it is the base of knowledge. Therefore, in that case the knowledge of the robot was highly reliable in order to answer questions during the human-robot interaction process. With respect to the colour extraction, we can see also good results despite the non favorable illumination and conditions of the environment. In addition, when an object does not have an uppermost colour or it is transparent, the sub-module of colour extraction does not work with precision-focus.

Table 4. Precision-Recall of the object classification, location and colour.

Obj. type	Classification		Location		Colour
	Prec.	Rec.	Prec.	Rec.	Acc.
Cup	100	100	100	100	33.33
Bottle	66.6	100	100	100	0
Laptop	100	100	100	100	100
Diningtable	100	100	100	100	0
Chair	100	75	100	100	66.66
Sofa	100	100	100	100	100
Book	100	100	100	100	100
Average	95.23	96.42	100	100	57.14

5.2 Human-Robot Interaction

To evaluate the dialogue module which manages the interaction with users, we have assessed two principal points. First, we evaluated the transcriptions generated from the original user sentence and the CFR outputs. In Table 5 there are some examples of typical interactions between the robot and a human, where is shown the dialogue and the CFRs generated. On average, SENNA with the rule-based semantic extractor was a very useful tool for obtaining always a valid CFR to understand what are the needs of the user and to consult its base of knowledge.

Table 5. Dialogue examples from the human-robot interaction.

Interaction	CFR
• HUMAN: Where is the sofa? • ROBOT: The blue sofa is in the living-room	BEING_LOCATED("theme" : "sofa")
• HUMAN: Can you look for a laptop? • ROBOT: I know 2 laptops, a black in the dinning-room and a gray in the living-room	SEARCHING("theme" : "laptop")
• HUMAN: The black one • ROBOT: All right, I will search the black laptop in the dinning-room	NO-VERB("theme" : "black")
• HUMAN: Can you bring me a sofa? • ROBOT: All right. I will bring the black sofa in the living-room	BRINGING("theme" : "sofa")
• HUMAN: Have you seen the blue cup? • ROBOT: The blue cup is in the living-room	BEING_LOCATED("theme" : "blue cup")
• HUMAN: Where are my glasses? • ROBOT: Sorry, I don't know any glasses	BEING_LOCATED("theme" : "glasses")

Second, we have analyzed the naturalness of the dialogue. To address that, ten different people has interacted with the robot and later, they have completed a test with satisfaction questions about the interaction. There were four questions in the test and all the answers were a range from 1 to 3. With these question, we tried to measure the accuracy of the modules. The questions were the following:

1. *How was the **naturalness** of the whole interaction?*, this question tries to measure the naturalness of the complete dialogue module, that is, the speech understanding sub-module, the dialogue control sub-module and the speech synthesis sub-module.

2. *How accurate was the* **speech understanding** *of the robot?*, this question measures the accuracy of the speech understanding sub-module.
3. *Are the robot's* **answers** *suitable?*, this question try to measure the dialogue control sub-module.
4. *Was the robot's* **knowledge** *of the environment accurate?*, this question measure the accuracy of the whole visual module.

The test results obtained are shown in Fig. 8, here we can observe that for 1 person (10%) the naturalness was *low*, for 5 persons (50%) was *medium* and for 4 persons (40%) was high, therefore the dialogue module is suitable but it could be improved. For the second and third questions, a majority (80% and 70% respectively) though that the understanding and the answers were appropriate, a minority (20%) considered that they were just correct and only one person (10%) admitted the answers were poor, it means that the speech understanding sub-module and the dialogue control sub-module are quite acceptable in isolation. The question about the knowledge obtained the worst results but it still was a good result, for 2 persons (20%) was *low*, for 6 persons (60%) was *medium* and for 2 persons (20%) was *high*, it means that the visual module had some mistakes in the recognition of the objects and its colours. Specifically, many persons argued that sometimes the colour that the robot had of several objects was not correct. On average, the users were satisfied with the interaction and they all maintained that the robot's knowledge of the objects on the environment was quite accurate.

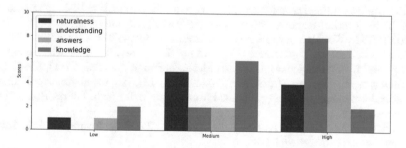

Fig. 8. Test results (Naturalness, understanding, answers and knowledge).

6   Conclusion and Future Work

We have presented the design and evaluation of a complete system to recognize objects from an indoor environment, storing that information in the knowledge base of the robot, and using it in the interaction with humans. The evaluation has been carried out in a use case that comprises several sub-tasks like object detection and classification, object localization and human-robot interaction. In view of the results obtained and considering that this is preliminary prototype of the final system, we can conclude that the recognition module of the objects shows promising results but the results regarding colour extraction need to be improved to be useful in the dialogue system. We can also conclude that the use of the knowledge in the interaction is adequate, and that the speech understanding system is quite accurate.

As future work, we plan to extend the developed system to carry out two more tasks. On one side, the robot should be able to alter its knowledge when it is interacting with humans. This means the robot will verify its knowledge with humans and it will have more reliable knowledge. In this way, the robot could ask the users about its uncertain knowledge to improve the object recognition, and even rectifying it while it is talking with an user.

On the other hand, the robot could be capable to track objects in the well-known environment to identify if the objects have been moved to other location or if they have disappeared from the scene.

Acknowledgements. This work has been partially sponsored by the Spanish Ministry of Economy and Competitiveness under grant number TIN2015-65686-C5-3-R and by the Regional Council of Education, Culture and Sports of Castilla-La Mancha under grant number SBPLY/17/180501/000493, supported with Feder funds.

References

1. Bandera, A., Bandera, J.P., Bustos, P., Férnandez, F., García-Olaya, A., García-Polo, J., García-Varea, I., Manso, L.J., Marfil, R., Martínez-Gómez, J., Núñez, P., Perez-Lorenzo, J.M., Reche-Lopez, P., Romero-González, C., Viciana-Abad, R.: LifeBots I: building the software infrastructure for supporting lifelong technologies. In: Third Iberian Robotics Conference (ROBOT), Sevilla, Spain (2017)
2. Bradski, G., Kaehler, A.: Opencv. Dr. Dobbs J. Softw. Tools **3** (2000)
3. Chan, M., Estève, D., Escriba, C., Campo, E.: A review of smart homespresent state and future challenges. Comput. Methods Progr. Biomed. **91**(1), 55–81 (2008)
4. Collobert, R., Weston, J., Bottou, L., Karlen, M., Kavukcuoglu, K., Kuksa, P.: Natural language processing (almost) from scratch. J. Mach. Learn. Res. **12**(Aug), 2493–2537 (2011)
5. Durette, P.N.: Google text-to-speech (version 2.0.1) [software] (2018). https://github.com/pndurette/gTTS
6. Goodrich, M.A., Schultz, A.C., et al.: Human–robot interaction: a survey. Found. Trends® Hum. Comput. Interact. **1**(3), 203–275 (2008)
7. Gross, H.-M., Schroeter, C., Mueller, S., Volkhardt, M., Einhorn, E., Bley, A., Martin, C., Langner, T., Merten, M.: Progress in developing a socially assistive mobile home robot companion for the elderly with mild cognitive impairment. In: 2011 IEEE/RSJ International Conference on Intelligent Robots and Systems (IROS), pp. 2430–2437. IEEE (2011)
8. Hartigan, J.A., Wong, M.A.: Algorithm as 136: a k-means clustering algorithm. J. Roy. Stat. Soc. Ser. C (Appl. Stat.) **28**(1), 100–108 (1979)
9. Iocchi, L., Kraetzschmar, G., Nardi, D., Lima, P.U., Miraldo, P., Bastianelli, E., Capobianco, R.: Rockin@home: Domestic robots challenge. In: RoCKIn, chap. 3. IntechOpen, Rijeka (2017)
10. Littlewort, G.C., Bartlett, M.S., Fasel, I.R., Chenu, J., Kanda, T., Ishiguro, H., Movellan, J.R.: Towards social robots: automatic evaluation of human-robot interaction by facial expression classification. In: Advances in Neural Information Processing Systems, pp. 1563–1570 (2004)

11. Redmon, J., Divvala, S., Girshick, R., Farhadi, A.: You only look once: unified, real-time object detection. In: Proceedings of the IEEE Conference on Computer Vision and Pattern Recognition, pp. 779–788 (2016)
12. SPARC Robotics: Robotics 2020 multi-annual roadmap for robotics in Europe. In: SPARC Robotics, EU-Robotics AISBL, The Hauge, The Netherlands (2016). Accessed 5 Feb 2018
13. Zeng, Z., Pantic, M., Roisman, G.I., Huang, T.S.: A survey of affect recognition methods: audio, visual, and spontaneous expressions. IEEE Trans. Pattern Anal. Mach. Intell. **31**(1), 39–58 (2009)
14. Zhang, A.: Speech recognition (version 3.8) [software] (2017). https://github.com/Uberi/speech_recognition

Adaptation of the Difficulty Level in an Infant-Robot Movement Contingency Study

José Carlos Pulido[1]([✉]), Rebecca Funke[2], Javier García[1], Beth A. Smith[2], and Maja Matarić[2]

[1] Universidad Carlos III de Madrid, Madrid, Spain
{jcpulido,fjgpolo}@inf.uc3m.es
[2] University of Southern California, Los Angeles, USA
{rfunke,beth.smith,mataric}@usc.edu

Abstract. This paper presents a personalized contingency feedback adaptation system that aims to encourage infants aged 6 to 8 months to gradually increase the peak acceleration of their leg movements. The ultimate challenge is to determine if a socially assistive humanoid robot can guide infant learning using contingent rewards, where the reward threshold is personalized for each infant using a reinforcement learning algorithm. The model learned from the data captured by wearable inertial sensors measuring infant leg movement accelerations in an earlier study. Each infant generated a unique model that determined the behavior of the robot. The presented results were obtained from the distributions of the participants' acceleration peaks and demonstrate that the resulting model is sensitive to the degree of differentiation among the participants; each participant (infant) should have his/her own learned policy.

Keywords: Socially assistive robotics · Infant-robot interaction
User adaptation · Reinforcement learning

1 Introduction

Infants produce a variety of movements in order to modulate task-specific actions such as reaching, crawling, and walking [1,2]. Through a dynamic process of exploration and discovery, they learn how to control their bodies and interact with their environments. In contrast to typically developing (TD) infants, infants at risk (AR) for developmental delays often have neuromotor impairments

This work was supported by NSF award 1706964 (PI: Smith, Co-PI: Matarić). In addition, this work was developed during an international mobility program at the University of Southern California being also partially funded by the European Union ECHORD++ project (FP7-ICT-601116), the LifeBots project (TIN2015-65686-C5) and THERAPIST project (TIN2012-38079).

© Springer Nature Switzerland AG 2019
R. Fuentetaja Pizán et al. (Eds.): WAF 2018, AISC 855, pp. 70–83, 2019.
https://doi.org/10.1007/978-3-319-99885-5_6

involving strength, proprioception, and coordination. These challenges can lead to greater difficulty with movement and potentially a decreased motivation to move and explore.

Past works have used wearable sensors and/or 3-dimensional motion analysis systems to assess differences in movement patterns between infants with TD and infants AR or with developmental delays. Studies have demonstrated that movement variables such as kicking frequency, spatiotemporal organization, and interjoint and interlimb coordination are different between infants with TD and infants AR [3], with intellectual disability [4], with myelomeningocele [5,6], with Down syndrome [7], or born preterm [8]. Studies have also shown that the acquisition of new motor skills is correlated to subsequent cognitive development in infancy [9,10], thus interventions to promote motor skills have the potential to be used to enhance the overall infant development.

In the first part of this contingency study, the goal was for infants to discover and learn that the movements of a humanoid robot are contingent upon their movement. The robot performed a reward action (kicking a ball on a string) contingently, in response to a desired movement by the infant. Specifically, the robot rewarded the infant when s/he produced a leg movement above a specified, constant acceleration value, which we call the activation threshold. In the second part of this contingency study, we created a personalized contingency feedback adaption system that aims to encourage infants to gradually increase their peak acceleration of each movement.

This paper focuses on the evaluation of a reinforcement learning (RL) algorithm that moderates the adaptation of the activation threshold using the data distributions of the acceleration peaks of every infant from the first part of the contingency study. The experimentation presented here uses those data as input for the model, to generate activation threshold values that adjust to each distribution individually. This proof-of-concept of the model is a necessary step before carrying out a study with infants.

This paper is structured as follows: Sect. 2 presents related work from multiple fields. Next, Sect. 3 explains the origin of the infants' data from the first part of the contingency study, summarizing the foundational study that was carried out. Section 4 provides the details of the proposed model from the second part of the contingency study, from the discretization to build the set of thresholds to the RL-based approach. Section 5 presents a simulation of the model using the infant data. Finally, Sect. 6 summarizes the work and outlines next steps of this research.

2   Related Work

This multidisciplinary project brings together and builds on insights from multiple research areas. Section 2.1 describes the basic theory of infant motor development and the basis of contingency studies. Section 2.2 describes the importance of early intervention in atypical motor development and the need for personalized adaptation for each infant.

2.1 Infant Motor Learning and Adaptation

Current developmental theory proposes that infants learn the connection between their body and the environment by making frequent exploratory movements that help them to develop task-specific actions [1,2]. For instance, when nine-month-old infants are placed in a jumper toy, they adjust the timing and force generation of their legs to optimize bouncing [11]. Our work used wearable inertial sensors attached to the infant's limbs to track the acceleration and angular velocity of each limb throughout the motor exploration task.

To motivate infant movements, researchers use contingency feedback paradigms. Historically, infant contingency studies used a mobile paradigm where a specific arm or leg is attached to the mobile with a string. The more the infant moves the attached limb, the more sound and motion are generated by the overhead mobile [12]. Contingency studies have demonstrated that, when movements are reinforced by mobile motion, infants with typical development as young as three months old can increase the movement rate of the arm [13], increase the kicking rate of the leg [14,15], move through a specific knee joint angle [16], produce more in-phase interlimb coordination by simultaneously moving both legs together [17], produce more in-phase hip-knee intralimb coordination by simultaneously extending the hip and knee of one leg [18], or produce selective hip-knee intralimb coordination (hip flexion with knee extension) by kicking a panel [19] or moving a foot vertically across a height threshold [20].

Those studies focused on reinforcing motion patterns; in this work we reinforce precise kinematic values, specifically the peak acceleration of a movement, aiming to encourage infants to increase the peak acceleration of their leg movements over time.

2.2 Infant Developmental Intervention

The main characteristic of this population is its enormous heterogeneity, since in such early stages, the aspects in the development and behavior patterns can vary enormously between individuals. That is why it is difficult to establish general guidelines and professionals need to make a more personalized analysis.

Approximately 9% of all infants in the United States are AR and could potentially benefit from early intervention services to address motor, cognitive, and/or social development [21]. All development domains, such as motor, cognitive, and social, are related, thus an intervention in one domain may provide benefits in all areas of development [15]. Despite this, the current standard of care for early intervention practice is to provide infrequent, low-intensity movement therapy or no intervention in infancy [22,23]. New research has shown that early, intense, and targeted therapy intervention has the potential to improve neurodevelopmental structure and function [24]. Despite this potential gain, it can be challenging to find feasible and resource-efficient ways to deliver this type of intervention in infancy. Our proposed solution is to use a non-contact *socially assistive* humanoid robot to provide demonstrations and feedback aimed to encourage infants in movement exploration tasks. A key aspect of the efficacy of this approach is the inclusion of *personalized models* appropriate for each infant participant that adapt the exploration task and difficulty to the specific infant, potentially allowing for higher engagement and improved learning.

Graded cueing is an approach that also aims at personalizing the level of task difficulty, by using increasingly specific cues or prompts given to the user [25]. This technique has been successful in rehabilitation of patients with brain injury and stoke, and has also been explored with socially assistive robots used with children with autism spectrum disorder in learning appropriate social skills [26,27]. The application of this technique consists of a set of steps that are applied sequentially. First, the therapist prompts the patient if the patient is having difficulties completing the assigned task. If, after a while, the patient continues to have difficulty, the therapist gives an increasingly specific cue, i.e., from a general verbal cue of patient's body posture to a more specific cue such as imitating patient's posture to help them to correct it. The purpose of using graded feedback is to encourage the patient to do most of the work on their own. The referenced past works address this problem by implementing models based on finite state machines or Markov Decision Processes. It has been shown to lead to more efficient learning and better learning outcomes.

This work follows a very similar concept. Different levels of difficulty are established and the participant starts at a low level. Difficulty levels are related to thresholds of acceleration peaks. The learning model must find the policy that allows to move between the different levels from the participant's progress while maximizing the received reward (average acceleration). The idea is to adjust the specificity of the learning task – creating movements with higher acceleration – by adapting the acceleration threshold required to receive the contingency reward based on the infant's past performance on the task.

3   Model Training Data

The training data used in this work were collected in a previous study. We summarize the data collection only briefly here.

Eight infants with TD between the ages of 6 and 8 months participated in a contingency feedback experiment in the Greater Los Angeles area. Only TD infants were recruited for this study as the first step was to enable the system to adapt to typical infant exploratory movement behavior.

The infant was placed in front of a NAO robot in a chair that allowed for full leg mobility, as shown in Fig. 1. The infant wore a head-mounted eye tracker. Opal inertial movement sensors [28] were affixed to each infant limb using cuffs with pockets. The sensors tracked the tri-axial acceleration and angular velocity of each limb.

For two minutes, the infant's baseline movement was measured. During that time, the robot remained inactive. After the baseline, the robot demonstrated the reward action three times. The action was a basic knee flexion kick at a ball on a string. After the demo, the contingency phase of the study ran for eight minutes. If the infant produced an acceleration from the right leg above a

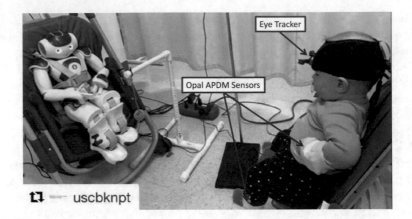

Fig. 1. An infant study participant interacting with the NAO robot in the previous study

fixed threshold of $3.0\,\mathrm{ms}^{-2}$, the robot performed the reward action. We chose the acceleration threshold based on a previous study that measured the accelerations of infant leg movements [29]. In this study, the difficulty of the activity did not change and the threshold remained fixed throughout the session. The study was approved by the University of Southern California Institutional Review Board under protocol #HS–14–00911.

Table 1 shows the acceleration peaks from the eight infants in the study. The variance among the participants is notable. The values of the means vary based on performance during the session. For instance, infant 1's mean peak acceleration is twice that of infant 5. Likewise, the maximum acceleration values reached by each infant and the number of acceleration peaks generated have a large variance. This is an indication that there is great heterogeneity in the participant pool, supporting personalized models rather than a generalized approach.

Table 1. Statistical outcomes of the study participants; N is the number of detected acceleration peaks for each participant.

VARIABLE	N	MEAN	STDEV	MIN	Q1	MEDIAN	Q3	MAX
ACC_PEAKS_U01	655	11.20	9.65	3.00	4.98	8.53	13.77	87.39
ACC_PEAKS_U02	417	9.77	8.06	3.01	4.30	6.31	12.51	45.66
ACC_PEAKS_U03	166	6.63	7.01	3.00	3.47	4.57	7.056	55.74
ACC_PEAKS_U04	326	9.51	8.61	3.02	4.21	5.87	11.15	63.49
ACC_PEAKS_U05	311	5.95	4.20	3.00	3.60	4.38	6.44	38.11
ACC_PEAKS_U06	499	8.98	8.69	3.00	4.20	5.78	9.38	72.41
ACC_PEAKS_U07	273	18.56	22.72	3.01	4.12	6.53	24.46	94.92
ACC_PEAKS_U08	359	7.11	6.16	3.01	3.85	4.98	7.88	48.26

The results of the previous study were promising and informed the objectives of this work. The majority of infants were able to learn the contingency with a set activation threshold. They moved above threshold more often in the contingency phase, in which they interacted with the robot, than in the baseline phase. Therefore, the next step is to try adjust the difficulty of the activity and determine if infants are able to adapt to a changing activation threshold.

4   User Adaptation Model

This section explains the proposed model for threshold adaptation in the infant movement contingency study. Section 4.1 provides a high level description of the problem. Section 4.2 explains the discretization of the peak acceleration values. Finally, Sect. 4.3 presents the RL approach for the adjustment of difficulty.

4.1   Problem Description

As noted earlier, the objective of the model is to adapt the activation threshold θ of the robot's reward action in real time. To achieve this, the contingency phase was segmented and the participants progress evaluated to determine the threshold for the next segment. Progress is defined in terms of the average of the acceleration peaks, since this work is focused on identifying thresholds that achieve a higher average in the acceleration of the infant's movements.

The threshold adaptation process was carried out during the contingency phase, in which the robot gave a reward (i.e., kicking the ball) each time the infant exceeded the current threshold, otherwise the robot remained still. Figure 2 is a representation of the contingency timeline divided into N segments. Each segment lasts 40 s; the duration was determined empirically to allow enough time for the infants to adapt to the new difficulty and for the model to receive enough learning experiences in every session.

The system started with an initial threshold θ^0 that changed over time based on the outcome obtained in each segment. At each time step n with $0 < n < N$, the model decides whether to raise, lower, or keep the threshold value θ^n, i.e., the difficulty of the activity (assuming higher thresholds are more difficult), based on the average value of the acceleration peaks obtained in the last segment. Each θ^n took its values from a set of thresholds Γ selected as described in Sect. 4.2.

The objective was to find the value of the threshold θ that maximized the acceleration of each infant's target limb. As shown in Sect. 3, the acceleration values reached by the infants are quite different from each other. Therefore, it is important to learn an individual model of each infant in order to obtain the threshold. The decision to modify the threshold is dependent on the threshold levels for each infant, the average acceleration value obtained in the previous segment, and the infant's degree of engagement. These variables were chosen because they are used by experts, and the aim is to learn a policy for each infant that adjusts the level of difficulty of the activity similar to the way a health care professional would.

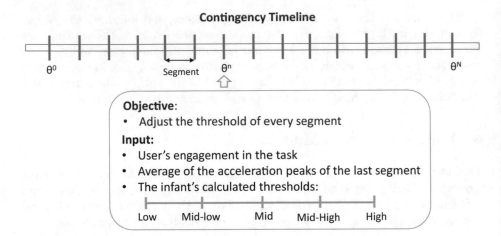

Fig. 2. Representation of the contingency problem

4.2 Discretization of the Acceleration Values

This section explains how the acceleration values of each infant were discretized to built a set $\Gamma = \{\theta_1, \theta_2, \ldots, \theta_q\}$ composed of q discretized threshold values that best match the data collected in their past sessions. In this study, 5 levels of difficulty related to acceleration peaks were established a priori, i.e., $q = 5$. Additionally, we assumed Γ is sorted in ascending order, i.e., $\forall i, j \ and \ i < j$, $\theta_i < \theta_j$ so that each threshold value corresponded to a level of difficulty: "low, mid–low, mid, mid–high, high".

As discussed in Sect. 3, preliminary analysis of the data revealed large differences in the movement data captured from the participating infants; some demonstrated double the average acceleration peaks of others. This evidence is consistent with previous research in development [30]. Together with potentially higher variability within and across infants in different AR populations, this determined the need to create independent models for each participant. This, in turn, suggested that each infant should have a discretized set of thresholds, Γ, adapted to their abilities.

Instead of using a uniform discretization, we used a *K-means* algorithm with $k = 5$ that allowed for finding the five centroids that best separated the acceleration data for each infant [31]. The centroids were directly related to the five levels of difficulty of the problem. Therefore, each threshold value $\theta_i \in \Gamma$ corresponded to a different centroid. Figure 3 shows an example for the data gathered from infant 1. The graph is the representation of the allocation of the instances to the different clusters found by the algorithm (the blue points corresponds to the instances in cluster 1, the green points to the instances in cluster 2, and so on). Furthermore, each cluster is represented by a centroid that corresponds to a value associated with the level of difficulty (in this case, $\Gamma = \{4.97, 10.81, 17.32, 28.89, 52.56\}$). In this example, and in most of the participants, there is no homogeneous allocation of the instances in the clusters due to the way in which the data are distributed: 47 % (low), 29 % (mid–low), 15 % (mid), 6% (mid–high), 2% (high) for the infant 1. This means that most instances are concentrated around low levels of acceleration, since infants reach the highest peaks of acceleration at specific times.

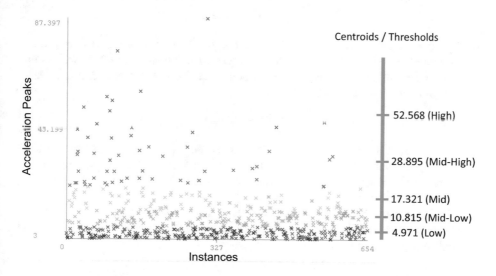

Fig. 3. Estimation of thresholds of the infant 1 using K-Means for the discretization of the accelerations peaks

4.3 Mapping the Threshold Adaptation Problem onto Reinforcement Learning

In this section, we describe the mapping of the problem of threshold adaptation of an infant described in Sect. 4.1 onto an RL approach. Such modeling requires defining all the elements of a Markov Decision Process (MDP): the state and action spaces and the reward and the transition functions [32]. We consider this to be an episodic task, where for each episode the infant is evaluated in N steps.

In this work, a state $s \in S$ is a tuple in the form $s^n =< \xi^n, \theta^n >$, where ξ^n and θ^n are respectively the disengagement of the infant and the current threshold of the system at step n. Feature ξ is a binary feature, i.e., $\xi \in \{0, 1\}$, where $\xi = 0$ if the infant is engaged, and $\xi = 1$ otherwise. Instead, feature θ takes values from the discrete set $\Gamma = \{\theta_1, \theta_2, \ldots, \theta_q\}$ built by discretizing the acceleration values of each infant, as described in Sect. 4.2. Therefore, the size of the state space S is $2 \times q$.

In state s^n, the agent performs an action $a^n \in A$. We consider the action space A as being composed of three actions, $A = \{-1, 0, 1\}$. These actions are used to decrease, leave as is, or increase, respectively, the threshold θ^n of the current state.

After performing an action a^n in state s^n, the agent transits to a new state $s^{n+1} = <\xi^{n+1}, \theta^{n+1}>$. A transition function is required to compute the values for ξ^{n+1} and θ^{n+1}. The value of ξ^{n+1} is computed using Eq. 1:

$$\xi^{n+1} = \begin{cases} 1, & \text{if } countHits < 2. \\ 0, & \text{otherwise.} \end{cases} \tag{1}$$

where $countHits$ is the number of times the infant moves with an acceleration above or below threshold θ^n in step n. To compute the value of θ^{n+1}, we assume that $\theta^n = \theta_i$, i.e., θ^n at step n corresponds with the i-th threshold in Γ. Then, we compute θ^{n+1} as in Eq. 2.

$$\theta^{n+1} = \theta_{i+a^n} \tag{2}$$

Therefore, if $a^n = 1$, the threshold is increased and θ^{n+1} takes the value of the $(i+1)$-th element in the Γ set, i.e., $\theta^{n+1} = \theta_{i+1}$. Conversely, if $a^n = -1$, the threshold is decremented and takes the value of the $(i-1)$-th element, i.e., $\theta^{n+1} = \theta_{i-1}$. If it is unchanged, then $\theta^{n+1} = \theta_i$.

Finally, when the learning agent performs an action a^n in a state s^n and moves to a state s^{n+1}, it also receives a reward signal r^n. We formulate the reward function as shown in Eq. 3.

$$r^n = \begin{cases} 0, & \text{if } countHits = 0. \\ avgSuccAcc \times (countSuccHits/countHits), & \text{otherwise.} \end{cases} \tag{3}$$

where $avgSuccAcc$ is the average acceleration of the infant's movements above threshold θ^n, $countSuccHits$ is the number of times the infant moves with an acceleration above the threshold θ^n, and $countHits$ is the number of times the infant moves (above or under the threshold θ^n). The rationale behind the reward function in Eq. 3 is as follows. If the infant does not move, the reward received is 0. If the infant moves ($countHits > 0$), and the threshold θ^n is exceeded ($countSuccessHits > 0$), the reward is greater than 0. If the threshold is easily exceeded by the infant, the reward is expected to be higher, consistent with a higher threshold. Conversely, if the threshold is not easily exceeded by the infant, the reward decreases, since $countSuccessHits$ tends to 0.

Finally, the reward function in Eq. 3 is different from the reward the robot provides to the infant. The former is used to learn a policy by RL to regulate the threshold θ that best fits the infant, while the latter is used to motivate the infant every time the infant exceeds the current threshold.

5   Simulation Evaluation of the Model

This section describes the evaluation of the model from Sect. 4. The objective is to extensively test the model prior to a study with infants by using the data from the first part of the contingency study described in Sect. 3. In a real scenario, the discretization of the acceleration values would be done individually from the data of the past sessions of each of the infants. To train the model, the peaks of infant movement acceleration were simulated and used as input to the RL model. Acceleration peaks of the participants typically followed an exponential distribution where there was a higher concentration of instances at low accelerations and fewer at high accelerations, as can be shown in Fig. 4. From the calculated distributions of each of the infants, the system generates random acceleration values that follow this distributions. In this way, it can be said that the behavior of every infant was being imitated, in terms of acceleration, based on their past experiences.

The objective of this simulation evaluation was to test the behavior of the model with two completely different infants: infant 5 and infant 7. According to Fig. 1, infant 5 obtained an average peak acceleration of 5.956 with a maximum value of 38.101, while infant 7 obtained an average peak acceleration of 18.56 with a maximum value of 94.92. Although they were very different, both followed an exponential distribution, as can be seen in Fig. 4. After applying the discretization described in Sect. 4.2, the set of thresholds for infant 5 were $\Gamma = \{3.41, 4.97, 7.88, 12.21, 20.87\}$ while the those for infant 7 were $\Gamma = \{3.82, 7.58, 16.55, 31.77, 56.99\}$. Both sets presented different values in line with the outcomes of each infant.

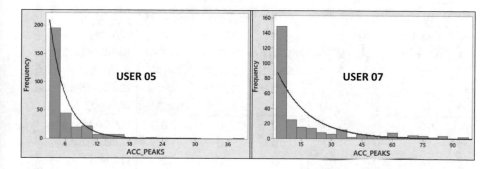

Fig. 4. Graphs of the distributions of the acceleration peaks of infants 5 and 7.

The simulation followed the approach presented in Fig. 2, in which the phase of contingency was divided into steps. Every step was an experience for the model, in which the values of acceleration peaks were created from the distribution of each infant, see Fig. 4.

For each infant, we simulated 50 episodes of 20 steps as described in Sect. 4.3. We used Q-Learning, and ϵ-greedy as exploration-exploitation strategy [32]. Table 2 shows the resulted Q-tables for infants 05 and 07 at the end of the learning process. Between state $S0$ and state $S4$ are the states when the infant was engaged with the task, i.e., $\xi = 0$, while from state $S5$ to state $S9$, the infant was disengaged, i.e., $\xi = 1$.

Significant differences can be seen between the Q-tables. The values of infant 7 are higher than those of infant 5, since the episodes generated with the first one contributed with greater rewards than those of the second. Looking at the highest value of each row, the policy learned can be found for each participant. For infant 5, the resulting policy considers more adequate to stay in a low threshold. For instance, if we consider that the infant starts in state $S0$, the best action is to stay in this state (the action *stay* is the action with the highest Q-value). However, if the infant starts in state $S3$ and is never disengaged, the policy decides to transit firs to $S2$, then to $S1$ and, finally, to $S0$, the state with the threshold that best suits the infant. As a last example, if we consider the infant starting in state $S4$, the system could transit to state $S8$ (i.e., system transits from a state with a high threshold to a state with a mid-high threshold, although the infant has disengaged). In $S8$, the best action is to reduce the threshold, so that the system could move to $S2$ (if we assume the infant is engaged again), then to $S1$, and finally to $S0$. Finally, it is important to note that the rows for the states $S5$ and $S6$ are all 0. This is intuitive since the infant is never disengaged when the system is in low or mid-low threshold values and, hence, the states $S5$ and $S6$ would be never visited.

Table 2. Results of Q tables of the simulated experiments of infants 05 and 07.

	USER 05			USER 07				
	UP	STAY	DOWN	UP	STAY	DOWN	(deng/th)	
S0	86.34	88.98	0	337.42	335.97	0	(0/Low)	Engaged
S1	84.09	87.76	89.30	350.06	348.44	345.34	(0/Mid-Low)	
S2	80.27	85.77	87.05	330.02	335.30	341.83	(0/Mid)	
S3	73.27	75.74	86.03	300.29	322.43	342.31	(0/Mid-High)	
S4	0	0	32.22	0	301.57	318.51	(0/High)	
S5	0	0	0	0	0	0	(1/Low)	Disengaged
S6	0	0	0	0	0	0	(1/Mid-Low)	
S7	0	61.03	0	0	0	0	(1/Mid)	
S8	68.47	78.44	83.34	0	0	0	(1/Mid-High)	
S9	0	70.93	78.50	0	275.99	313.25	(1/High)	

Instead, infant 7 was able to get higher acceleration values between mid–low to mid thresholds, since s/he had higher accelerated movements in past sessions. Following the same reasoning as in the other Q-table, if we consider the infant that starts in state $S0$ and is never disengaged, the system first transits to $S1$, and then to $S2$. Then, a loop occurs: in $S3$ it decides to reduce the threshold and transits to $S2$. Therefore, the threshold that best suits this infant is between the states $S2$ and $S1$. Finally, as in the previous case, the rows for states $S5$ to $S8$ are 0, as these states are never visited; this infant does not disengaged until reaching a high threshold value in state $S9$.

6   Conclusion

This paper presented an approach for using a personalized reinforcement learning algorithm for infants learning to reach target leg movement acceleration. The RL-based model was able to determine the best threshold configuration in terms of peak acceleration. The results of the simulation were very promising; the model was sensitive to the high variance among the infant study participants. The policy learned for each participant indicated the thresholds that would reach higher rewards values. Since the reward function was related to the average of the acceleration peaks and the number of peaks detected, maintaining these thresholds in a session would help to maximize these two variables.

In a real infant-robot interaction scenario, higher difficulty levels would offer better rewards from the robot. Thus, the ultimate goal of this study is to determine whether the robot is able to encourage the infant to reach higher accelerations from their movements to get better rewards from the robot. This work validates the proof-of-concept of the model, making it ready for implementation in our upcoming contingency study of infant-robot interaction.

This novel work in socially assistive robotics for infant movement therapy is the basis for the upcoming studies that will extend the presented results. We plan to explore new reward functions that reinforce other aspects of the movement or allow the dissociation of one limb from the other. Additionally, we intend to integrate this socially assistive robot system into the next infant-robot contingency study to determine if the model helps with the adaptation of the infants achieving better results than with approaches based on fixed activation thresholds.

References

1. Gibson, E.J., Pick, A.D.: An Ecological Approach to Perceptual Learning and Development. Oxford University Press, Oxford (2000)
2. Thelen, E., Smith, L.: A Dynamic Systems Approach to the Development of Cognition and Action. The MIT Press, Cambridge (1994)
3. Smith, B., Vanderbilt, D.L., Applequist, B., Kyvelidou, A.: Sample entropy identifies differences in spontaneous leg movement behavior between infants with typical development and infants at risk of developmental delay **5**, 55 (2017)
4. Kouwaki, M., Yokochi, M., Kamiya, T., Yokochi, K.: Spontaneous movements in the supine position of preterm infants with intellectual disability. Brain Dev. **36**(7), 572–577 (2014)
5. Rademacher, N., Black, D.P., Ulrich, B.D.: Early spontaneous leg movements in infants born with and without myelomeningocele. Pediatric Phys. Ther. **20**(2), 137–145 (2008)
6. Smith, B.A., Teulier, C., Sansom, J., Stergiou, N., Ulrich, B.D.: Approximate entropy values demonstrate impaired neuromotor control of spontaneous leg activity in infants with myelomeningocele. Pediatr. Phys. Ther. **23**(3), 241–247 (2008)
7. McKay, S.M., Angulo-Barroso, R.M.: Longitudinal assessment of leg motor activity and sleep patterns in infants with and without down syndrome. Infant Behav. Dev. **29**(2), 153–168 (2006)

8. Geerdink, J.J., Hopkins, B., Beek, W.J., Heriza, C.B.: The organization of leg movements in preterm and full-term infants after term age. Dev. Psychobiol. **29**(4), 335–351 (1996)
9. Kermoian, R., Campos, J.: Locomotor experience: a facilitor of spatial cognitive development. Child Dev. **59**, 908–917 (1998)
10. Oudgenoeg-Paz, O., Volman, M.: Attainment of sitting and walking predicts development of productive vocabulary between ages 16 and 28 months. Infant Behav. Dev. **35**, 733–736 (1998)
11. Goldfield, E.C., Kay, B.A., Warren, W.H.: Infant bouncing: the assembly and tuning of action systems. Child Dev. **64**(4), 1128–1142 (1993)
12. Rovee-Collier, C.K., Gekoski, M.J.: The economics of infancy: a review of conjugate reinforcement. In: Advances in Child Development and Behavior, vol. 13, pp. 195–255. Elsevier (1979)
13. Watanabe, H., Taga, G.: General to specific development of movement patterns and memory for contingency between actions and events in young infants. Infant Behav. Dev. **29**(3), 402–422 (2006)
14. Heathcock, J.C., Bhat, A.N., Lobo, M.A., Galloway, J.: The performance of infants born preterm and full-term in the mobile paradigm: learning and memory. Phys. Ther. **84**(9), 808–821 (2004)
15. Lobo, M.A., Galloway, J.C.: Assessment and stability of early learning abilities in preterm and full-term infants across the first two years of life. Res. Dev. Disabil. **34**(5), 1721–1730 (2013)
16. Angulo-Kinzler, R.M., Ulrich, B., Thelen, E.: Three-month-old infants can select specific leg motor solutions. Motor Control **6**(1), 52–68 (2002)
17. Thelen, E.: Three-month-old infants can learn task-specific patterns of interlimb coordination. Psychol. Sci. **5**(5), 280–285 (1994)
18. Angulo-Kinzler, R.M.: Exploration and selection of intralimb coordination patterns in 3-month-old infants. J. Motor Behav. **33**(4), 363–376 (2001)
19. Chen, Y.-P., Fetters, L., Holt, K.G., Saltzman, E.: Making the mobile move: constraining task and environment. Infant Behav. Dev. **25**(2), 195–220 (2002)
20. Sargent, B., Schweighofer, N., Kubo, M., Fetters, L.: Infant exploratory learning: influence on leg joint coordination. PLoS ONE **9**(3), e91500 (2014)
21. Rosenberg, S.A., Robinson, C.C., Shaw, E.F., Ellison, M.C.: Part c early intervention for infants and toddlers: percentage eligible versus served. Pediatrics **131**(1), 38–46 (2013)
22. Roberts, G., Howard, K., Spittle, A.J., Brown, N.C., Anderson, P.J., Doyle, L.W.: Rates of early intervention services in very preterm children with developmental disabilities at age 2 years. J. Paediatr. Child Health **44**(5), 276–280 (2008)
23. Tang, B.G., Feldman, H.M., Huffman, L.C., Kagawa, K.J., Gould, J.B.: Missed opportunities in the referral of high-risk infants to early intervention. In: Pediatrics peds–2011 (2012)
24. Holt, R.L., Mikati, M.A.: Care for child development: basic science rationale and effects of interventions. Pediatr. Neurol. **44**(4), 239–253 (2011)
25. Bottari, C., Dassa, C., Rainville, C., Dutil, E.: The IADL profile: development, content validity, intra- and interrater agreement. Can. J. Occup. Ther. **77**(2), 345–356 (2009)
26. Feil-Seifer, D., Matarić, M.: A simon-says robot providing autonomous imitation feedback using graded cueing, In: Poster paper in International Meeting for Autism Research (IMFAR) (2012)

27. Greczek, J., Kaszubski, E., Atrash, A., Matarić, M.: Graded cueing feedback in robot-mediated imitation practice for children with autism spectrum disorders. In: IEEE International Symposium on Robot and Human Interactive Communication (RO-MAN), pp. 561–566 (2014)
28. APDM Wearable Technologies, Portland, OR, USA, Opals. https://www.apdm.com/wearable-sensors/. Accessed 15 July 2018
29. Trujillo-Priego, I.A., Smith, B.A.: Kinematic characteristics of infant leg movements produced across a full day. J. Rehabil. Assist. Technol. Eng. **4**, 2055668317717461 (2017)
30. Adolph, K.E., Robinson, S.R.: Sampling development. J. Cogn. Dev. **12**(4), 411–423 (2011). https://doi.org/10.1080/15248372.2011.608190
31. Hartigan, J.A., Wong, M.A.: Algorithm as 136: a k-means clustering algorithm. J. Roy. Stat. Soc. Ser. C (Appl. Stat.) **28**(1), 100–108 (1979)
32. Sutton, R.S., Barto, A.G.: Reinforcement Learning 1: Introduction. MIT Press, Cambridge (1998)

Computer Vision and Robotics

Convolutional Neural Network vs Traditional Methods for Offline Recognition of Handwritten Digits

Edwin Antonio Enriquez[1], Nelly Gordillo[1], Luis Miguel Bergasa[2(✉)],
Eduardo Romera[2], and Carlos Gómez Huélamo[2]

[1] Department of Electrical and Computer Engineering,
Institute of Engineering and Technology,
Autonomous University of Ciudad Juarez, Juarez, Chihuahua, Mexico
al150671@alumnos.uacj.mx, nelly.gordillo@uacj.mx
[2] Electronics Department, University of Alcalá (UAH), Alcalá de Henares, Spain
{luism.bergasa,eduardo.romera}@uah.es, carlos.gomezh@edu.uah.es

Abstract. This paper compares Convolutional Neural Networks vs traditional features extraction and classification techniques for an offline recognition of handwritten digits application. The studied classification techniques are: k-NN, Mahalanobis distance and Support Vector Machines (SVM); and the hand-designed features extraction ones are: Hu Invariant Moments, Fourier Descriptors, Projections Histograms, Horizontal Cell Projections, Local Line Fitting and Zoning. The study was conducted in a practical application as is the validation of democratic elections using ballots of electoral scrutiny with non-homogeneous background. To do that it was necessary to use different preprocessing techniques (RGB conversion to gray scale, binarization and noise reduction) as well as a segmentation stage.

Keywords: Handwritten digits recognition
Hand-designed features extraction · Classification · CNN

1 Introduction

Offline handwritten character recognition has been an interesting topic in the fields of pattern recognition and deep learning from decades. As handwritten characters are unconstrained and topologically diverse, its recognition with pure offline information is not trivial. In an offline handwritten recognition system the writing is usually captured optically by a scanner and provided as an input image. Its goal is to digitize the handwriting by converting it into machine readable ASCII format [1]. Online methods have shown to be superior to offline ones in recognizing handwritten characters, but as several applications require optical recognition, offline methods continue being an active research topic.

In the electoral process of most of the developing countries, the scrutiny of the votes is manually done. In consequence, this electoral phase is prone to counting

R. Fuentetaja Pizán et al. (Eds.): WAF 2018, AISC 855, pp. 87–99, 2019.
https://doi.org/10.1007/978-3-319-99885-5_7

errors. Figure 1 depicts an example of a ballot corresponding to an electoral process in Mexico. This is an application where an optical handwritten digits recognizer can help to automate the counting process, minimizing the human errors. It should be noted that the electoral ballot is self-designed, therefore, it is not a predefined document where standard techniques can be used without a preprocessing and segmentation stage.

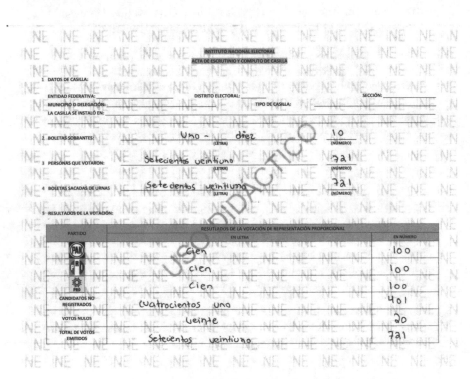

Fig. 1. Example of scrutiny ballot

In the state of the art there are many papers that have tackled this problem by using different traditional methods [2–4]. A survey of these methods can be found in [1]. In the last years, Convolutional Neural Networks (CNNs) [5] have received a lot of attention due to its ability to recognize handwritten digits, reaching a moderate good performance [6,7].

This paper compares some traditional methods versus a CNN for offline handwritten digits recognition in a real application of electoral ballot counting in Mexico. We have used the CHARS74k and MNIST datasets for training and a particular dataset called UACJ280, formed for 6904 digits extracted from real ballots, for testing. The paper is organized as follows: chapter two presents the foundations of used traditional methods of the state of the art as well as the used CNN for our handwritten digits recognition system. Later in chapter three the different architectures tested are detailed. Finally, chapter four presents some comparative experimental results obtained for the tested architectures and chapter five describes the conclusions of this work.

2   Foundations

2.1   Traditional Methods

Traditional handwriting recognition methods normally have four stages named: preprocessing, segmentation, feature extraction and classification. Hereafter, we sketch the foundations of the different methods used for each stage in our application.

Preprocessing. In this stage we apply a series of operations over the scanned image to improve its quality for the next segmentation stage. Firstly, the color image is resized and converted to a grayscale image, and after that the following techniques are applied: binarization and noise reduction.

Segmentation. A decomposition of the whole image into multiple segments is carried out in this stage. We use a simple but efficient simple pixel counting method for segmentation. The binarized image is scanned from left to right and from top to bottom. Then, an horizontal and vertical projection profiles are obtained. From these profiles lines and digits can be easily segmented by cropping the image at the minimum of the horizontal and vertical profiles respectively [1].

Feature Extraction. This stage finds out some representative hand-designed features for each digit useful for the next classification stage. The different used methods are shown below.

A very established shape features are the **Hu invariants moments** [8]. The 2-D moment of $(p + q)$ order of a $f(x, y)$ image, where the point x defines the columns and y the rows, is defined by:

$$m_{pq} = \sum_x \sum_y x^p y^q f(x, y) \tag{1}$$

For $p, q = 0, 1, 2, ...$, where the summations move through the spatial (x, y) coordinates of the image. Its corresponding central moment is:

$$\mu_{pq} = \sum_x \sum_y (x - \bar{x})^p (y - \bar{y})^q f(x, y) \tag{2}$$

where:

$$\bar{x} = \frac{m_{10}}{m_{00}} \tag{3}$$

$$\bar{y} = \frac{m_{01}}{m_{00}} \tag{4}$$

Being $(\bar{x}, \bar{y})$ the centroid of the digit. The normalized moments of order $(p+q)$ are:

$$\eta_{pq} = \frac{\mu_{pq}}{\mu_{00}^{\gamma}} \tag{5}$$

where:

$$\gamma = \frac{p+q}{2} + 1 \tag{6}$$

the γ function is only defined to $p+q = 2, 3, \ldots$ The set of seven Hu invariant moments [9] are insensitive to translation, rotation and scale. The invariant moments are expressed as follows:

$$\theta_1 = \eta_{20} + \eta_{02} \tag{7}$$

$$\theta_2 = (\eta_{20} - \eta_{02})^2 + 4\eta_{11}^2 \tag{8}$$

$$\theta_3 = (\eta_{30} - 3\eta_{12})^2 + (3\eta_{21} - \eta_{03})^2 \tag{9}$$

$$\theta_4 = (\eta_{30} + \eta_{12})^2 + (\eta_{21} + \eta_{03})^2 \tag{10}$$

$$\theta_5 = (\eta_{30} - 3\eta_{12})(\eta_{30} + \eta_{12})[(\eta_{30} + \eta_{12})^2 - 3(\eta_{21} + \eta_{03})^2] + (3\eta_{21} - \eta_{03})(\eta_{21} + \eta_{03})[3(\eta_{30} + \eta_{12})^2 - (\eta_{21} + \eta_{03})^2] \tag{11}$$

$$\theta_6 = (\eta_{20} - \eta_{02})[(\eta_{30} + \eta_{12})^2 - (\eta_{21} + \eta_{03})^2] + 4\eta_{11}(\eta_{30} + \eta_{12})(\eta_{21} + \eta_{03}) \tag{12}$$

$$\theta_7 = (3\eta_{21} - \eta_{03})(\eta_{30} + \eta_{12})[(\eta_{30} + \eta_{12})^2 - 3(\eta_{21} + \eta_{03})^2] + (3\eta_{12} - \eta_{30})(\eta_{21} + \eta_{03})[3(\eta_{30} + \eta_{12})^2 - (\eta_{21} + \eta_{03})^2] \tag{13}$$

Another method is the **Fourier Descriptors** [8,10], which use the coordinates of the image contours in the form of complex number $s(k)$ expressed as:

$$s(k) = x(k) + jy(k) \tag{14}$$

where $k = 0, 1, \ldots, K - 1$. To obtain the Fourier Descriptors $a(u)$ we use the Discrete Fourier Transform defined by:

$$a(u) = \sum_{k=0}^{K-1} s(k) \exp\left(-\frac{i2\pi uk}{K}\right) \tag{15}$$

On other side, [10] shows details of the **Projection Histograms** method expressed by:

$$P_{hor}(y_0) = \sum_{x=1}^{N} f(x, y_0) to 1 \le y_0 \le m \tag{16}$$

$$P_{ver}(x_0) = \sum_{y=1}^{M} f(x_0, y) to 1 \le x_0 \le n \tag{17}$$

where m and n are height and width of the image respectively.

By Zahid [11], the **Horizontal Cell Projections** method performs a partition of the character image in k regions. For the case of a horizontal cell projection, the characteristic vector of the r-th cell, corresponding to a character of $m \times n$ pixels, is expressed as $P_r = \langle P_1, P_2, ..., P_M \rangle$ in which:

$$P_i = \bigcup_{j=1}^{n/k} f\left(i, \frac{n(r-1)}{k} + j\right) \tag{18}$$

The characteristic vector of the image will be:

$$V = P_1 \cup P_2 \cup \dots P_k \tag{19}$$

A method proposed by Pérez et al. [12] is the **Local Line Fitting**. It consists of dividing the input image (character) into k cells (meshing) and obtaining three characteristics for each of them. Firstly, the number of pixels in 1's for each cell i belonging to a character is calculated (n_i) and then its value is divided between the total number of pixels that compose the character (N), that is:

$$f_{i1} = \frac{n_i}{N} \tag{20}$$

The second and third attribute consists of performing a orthogonal regression for each cell to obtain an estimated straight line for a set of data of the form $y = a_i + b_i x$, where a_i is the intersection with y axis and b_i is the slope of the straight line. The attributes f_{i2} and f_{i3} are:

$$f_{i2} = \frac{2b_i}{1 + b_i^2} \tag{21}$$

$$f_{i3} = \frac{1 - b_i^2}{1 + b_i^2} \tag{22}$$

In [13, 14] the **Zoning** method is presented, which divides the grayscale or binarized character in a $m \times n$ mesh. Each cell counts the number of pixels with 1 value (or the grayscale average) providing a vector of features $m \times n$ length.

Clasiffiers

- **k-Nearest Neighbor (k-NN):** This technique is one of the best known for classification. The input consists of the k closest training examples in the feature space. The output is a class membership. An object is classified by a majority vote of its neighbors, with the object being assigned to the class most common among its k nearest neighbors. When $k = 1$ k-NN becomes in a metric Euclidean distance classifier.
- **Mahalanobis distance:** This gives a measure of the distance between a features vector (x_{test}) and a distribution given by its mean vector (m_l), which represents the centroid of each class previously calculated in the training phase. This distance is similar to the Euclidean distance but in this case it uses a covariance matrix for each class. The covariance S is a quadratic matrix $n \times n$ length, whose main diagonal contains the variances of each feature $(S_{ij}^2 \mid i = j)$ and the rest of elements $i \neq j$ define the respective covariances values among features $(S_{ij}^2 \mid i \neq j)$.
- **Support Vector Machines (SVM):** It has been extensively used due to its easy training and the high performance obtained in classification. SVM is based on the choice of a kernel function and the penalty parameter C. The kernel maps nonlinear samples into a higher dimensional space and handles the nonlinear relations between class labels and features. SVM is widely used in classic machine learning where feature extraction phase is handmade and classification phase is learned from data in a previous training process.

2.2   Convolutional Neural Networks

CNNs are a class of deep, feed-forward artificial neural networks that has successfully been applied to image processing. They differ of traditional methods in that both the features extraction phase and the classification one are carried out in a previous deep learning process. CNNs use relatively little pre-processing compared to traditional image classification algorithms. This means that the CNN learns the filters that in traditional algorithms were hand-engineered. This independence from prior knowledge and human effort in feature design is its major advantage. CNN is a multi-layer network alternating operations of convolution and sub-sampling (features extraction) as well as full connexion and Gaussian connexion (classification). Each convolutional layer apply a convolution operation to the input image, passing the result to the next layer. Each convolutional neuron processes data only for its receptive field. Many CNN models can be found in the literature for classification: AlexNet [15], VGG [16], Le-Net [17], etc.

In this project we use a Le-Net-5 architecture because, according to [17], it is suitable for handwriting recognition. We keep the same pre-processing and segmentation stages that in the traditional methods but the features extraction and classification stages are carried out through this CNN.

3   System Architecture

This section shows the main characteristics of the implemented architectures according to the different stages explained before.

3.1   Pre-processing

The images to analyze are the areas where the counts of votes are located, as we depicts in the Fig. 2. As you can see, these areas are boxes present in the electoral scrutiny form to be fill and they have non homogeneous background because it usually includes an organization in charge of the process seal and watermarks. Pre-processing includes an image size reduction using bicubic interpolation, a conversion of RGB image to grayscale, a binarization and some noise reduction operations.

- **Binarization:** Consists of converting a grayscale image with pixel values between 0 and 255 to a binary image with $[0, 1]$ values using Otsu algorithm adaptive threshold [18]. In this case 0 indicates background pixels (black) while 1 indicates foreground pixels (white).
- **Noise Reduction:** This operation removes any unwanted bit-patterns, which are noise for the digits segmentation. To do that we apply morphological operators and small blobs removing. In a first instance an erosion of the binary image is carried out to filter out the digits to be recognized. Subsequently a dilatation is made to consolidate the remain blobs [10]. Finally, we eliminate the objects whose area is lower that 20% of the largest blob (digit).

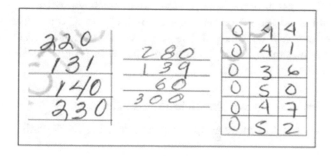

Fig. 2. Some images used

3.2   Segmentation

For the segmentation we apply a two steps process: lines segmentation and characters segmentation, based on pixel counting technique [1]. In the first step, a horizontal projection of the pre-processed binary image is made. We perform a row-by-row scan to detect a row of the image which projection value is zero. This means that the end of an information line is detected and therefore is segmented.

In the second step a vertical projection is applied. A column-by-column scan is performed looking for zero-values. These indicate the transitions among characters and are used to segment them. This process is repeated until the whole digits of the image are segmented.

3.3   Features Extraction Using Traditional Methods

At this stage, some hand-made parameters, which provide an accurate representation of the digits to the classification stage, are found out. The used techniques are shown below.

- **Hu Invariants Moments:** We use the seven invariant moments described in previous section, applying Eqs. (7)–(13).
- **Fourier Descriptors:** From the characteristics obtained by Eq. (15), only the first thirty-two descriptors of each digit are taking into account for the classification stage. This value was chosen in an experimental way.
- **Projection Histograms:** We use features calculated from Eqs. (16) and (17), separately and jointly, for classification. More details will be given in next section.
- **Horizontal Cell Projections:** For features extraction the digit patch is divided into three up to eight cells to obtain the horizontal and vertical projections. In the same way as the ordinary projections seen in the previous method, a separate comparison is made for the obtained horizontal and vertical projections.
- **Local Line Fitting:** Each digit patch is divided into a mesh of $n \times m$ cells. For each cell the features calculated by Eqs. (20)–(22) are extracted with the difference that a regression is done by the method of least squares to obtain the slope. The values of n and m are the same which comprise $n, m = 3, 4, ..., 8$.
- **Zoning:** This method, comparing the previous one, only divides the digit into a mesh of $n \times m$ cells for the same values seen previously and makes for each cell a count of the pixels belonging to the digit creating a vector of characteristics $n \times m$ length.

3.4   Classification Using Traditional Methods

In this stage each digit patch is assigned to a class using some traditional methods based on distance.

- **K-Nearest Neighbor:** This method calculate the distance of an unknown extracted features vector to the K nearest neighbors and classify it as belonging to the class to most of the neighbors belong. K was chosen in an experimental way to 15. There is a training phase consisting in obtaining a $k \times l$ matrix, where k are the chosen features and l is the number of characters used.
- **Mahalanobis Distance:** It is similar to the previous method but in this case the used distance metric takes into account the covariance of each class. In the previous training process the mean and covariance matrix must be calculated for each digit class.

- **SVM:** In this case we use a multi-class SVM classifier. We made several tests and the best results were obtained for an linear kernel and a penalty parameter $C = 1$.

3.5   CNN for Features Extraction and Classification

For the features extraction and classification stages, we have used a CNN based on LeNet-5 [17]. The architecture of our network is as follows:

- Layer S1 is an input patch corresponding to a digit of size 28×28 pixels.
- Layer C1 is the first convolutional layer with 20 feature maps of size 28×28. Each unit is connected to a 5×5 neighborhood of the input layer S1.
- Layer S2 is the second sub-sampling layer with 20 feature maps of size 14×14. Each unit is connected to a 2×2 neighborhood in the corresponding feature map in layer C1.
- Layer C3 is the second convolutional layer. This contains 50 feature maps of size 8×8. Each unit is connected to a 5×5 neighborhood in the corresponding feature map in layer S2.
- Layer S4 is the third sub-sampling layer containing 50 feature maps of 4×4 where each unit is connected to a 2×2 neighborhood of the layer C3.
- Layer C5 is the third convolutional layer. This contains 500 feature maps of size 1×1. Each feature map is connected to all 50 feature maps of layer S4. We have added a ReLu activation function that adds nonlinearity to the network.
- Layer F6 contains 60 units and it is fully connected to layer C5.
- Layer F7 (the output layer) is composed of Euclidean RBF units and it is a fully connected to layer F6. We have used a softmax layer, in which it shows the probabilities of each class, for error calculation.

4   Results

4.1   Handwritten Databases

For this work, the used training databases were CHARS74K [19] and MNIST [20]. The first database contains a total of 3,410 elements between the ten digits of 0–9, twenty-six characters from a to z in lowercase and another group of twenty-six characters but in uppercase. Each character has fifty-five types of writing. It should be noted that for the training in the algorithms implemented only the 0–9 digits were used, giving a total training of 550 images resized to a 28×28 resolution.

On the other hand, for the MNIST database, only the 60,000 digit training database was used, which was divided into two groups. The first consisted in a training group of 10,000 (MNIST1000) divided equally among the ten digits. The same was done with the second group of 54,000 elements (MNIST5400). Image dimensions were 28×28 for the distance algorithms and the convolutional network.

For the test phase the digits extracted from the ballots were used as a result of the pre-processing that will be detailed below. The UACJ280 database has a total of 6904 digits from 0 to 9. The dimensions for this database were the same that in the training phase.

Table 1. Classification percentages using K-Nearest Neighbor

Feature	CHARS74K	MNIST1000	MNIST5400
Hu Invariants Moments	41%	38%	40%
Fourier Descriptors	46%	53%	56%
Projections Histograms	63%	63%	66%
Cells Projections	**82%**	91%	94%
Local Line Fitting	78%	88%	92%
Zoning	80%	**92%**	**95%**

In the following tables, classification percentages for the UACJ280 database are shown as a function of the three different training databases and for the six features extraction techniques used for each classifier (k-NN, Mahalanobis distance and SVM). Table 1 shows the highest recognition percentages for the k-NN classifier. The Hu Invariant Moments presented the percentages of classification of smaller magnitude in comparison of the other features extraction algorithms. Results do not exceed 41% of correct detection for the UACJ280 database. On the other hand, the Fourier Descriptors showed variability in the classifications for each training set, not exceeding 56% of the total characters for the MNIST5400 database. The percentages of Projections Histograms are referring to those of horizontal type achieving recognition numbers between 63% and 66%. In the case of Cell Projections for five horizontal cells, it shows percentages higher than 80%, obtaining the highest percentage for the MNIST5400 database with more than 94% of correct detections. For the case of Local Line Fitting with a 6×6 grid, using the MNIST5400 as training dataset, 92% was obtained in recognizing the digits of UACJ280. Finally, for the method of Zoning, a similar behavior is shown as for the cell projections one.

Mahalanobis distance classifier (Table 2) shows different percentages and in most of the cases are lower compared to the previous classifier. CHARS74K shows the lowest percentages in recognition having about 41%. MNIST5400 is about 55% and finally MNIST5400 is about 62%. It should be noted that percentages shown in this table are obtained for the following features extractor cases: Projection Histograms uses only the horizontal type. Cell Projections applies a combined evaluation between horizontal and vertical projections of three cells. Local Line Fitting uses a 3×3 grid while Zoning uses a 5×5 grid.

For the evaluation applying SVM (Table 3) there are percentages of recognition below that expected maybe due to only a linear kernel was used. It should be noted that the percentages of classification for Hu Invariant Moments are the

Table 2. Classification percentages using Mahalanobis distance

Feature	CHARS74K	MNIST1000	MNIST5400
Hu Invariants Moments	41%	41%	43%
Fourier Descriptors	37%	37%	37%
Projections Histograms	40%	55%	60%
Cells Projections	30%	73%	81%
Local Line Fitting	21%	34%	57%
Zoning	**76%**	**89%**	**93%**

lowest (same characteristic in other classifiers) between 12% and 40%. In contrast, for Fourier Descriptors the percentages are between 40% and 50% between the three bases of training where 47% of the UACJ280 test elements were classified by MNIST1000. Highlight the Histograms of Horizontal Projections have 60% expected value in conjunction with the three training databases described above. Meanwhile, projections by eight cells have recognition percentages of 86% and 89% for the MNIST1000 and MNIST5400 databases, respectively. Results for CHARS74K database show a high difference of 22% compared to MNIST1000. The Local Line Fitting method using a 6×6 grid presents a behavior similar to the previous method, obtaining a difference of 2% between their respective training bases. Finally, the extraction by zoning method for 5×5 grid presents 68% for CHARS74K, 85% for MNIST1000 and 87% for MNIST5400.

Table 3. Classification percentages using support vector machines

Feature	CHARS74K	MNIST1000	MNIST5400
Hu Invariants Moments	12%	39%	40%
Fourier Descriptors	40%	47%	50%
Projections Histograms	55%	61%	63%
Cells Projections	64%	**86%**	**89%**
Local Line Fitting	66%	84%	87%
Zoning	**68%**	85%	87%

Results for the classification technique are shown in Table 4. Convolutional Neural Networks were analyzed with fifteen and twenty-five training periods for the neural network. The classification percentages obtained by this technique were the highest in comparison to the traditional classification techniques. In the case of fifteen seasons, 85% classification percentage is obtained for CHARS74K; 93% for MNIST1000 and 98% for MNIST5400, which is 5% over MNSIT1000. For the case for 25 epoch, the same trend is observed, that is, for CHARS74K and MNIST1000 datasets there is a difference of 1% over results with fifteen seasons, and for MNIST5400 the same percentage prevails.

Table 4. Classification percentages using Convolutional Neural Networks

EPOCH	CHARS74K	MNIST1000	MNIST5400
15	85%	93%	98%
25	**86%**	**94%**	**98%**

5 Conclusions

We conducted a comparative study of Convolutional Neural Networks vs. traditional methods of classification, which led to a practical level that is the classification for an offline recognition of hand written digits application as is the validation of democratic election using ballots of electoral scrutiny with non homogeneous background. We conclude that results obtained with the CNN are quite better than the obtained with traditional methods. It should be noted that all the classifiers were susceptible to the amount of training data, in other words, the larger the database to be trained, the better the classification of the test elements.

Acknowledgment. This work has been partially funded by the Spanish MINECO/ FEDER through the SmartElderlyCar project (TRA2015-70501-C2-1-R), the DGT through the SERMON project (SPIP2017-02305), and from the RoboCity2030-III-CM project (Robótica aplicada a la mejora de la calidad de vida de los ciudadanos, fase III; S2013/MIT-2748), funded by Programas de actividades I+D (CAM) and cofunded by EU Structural Funds.

References

1. Manoj, A., Borate, P., Jain, P., Sanas, V., Pashte, R.: A survey on offline handwriting recognition systems (2016)
2. Parizeau, M., Lemieux, A., Gagné, C.: Character recognition experiments using Unipen data. In: ICDAR, p. 0481. IEEE (2001)
3. Ghosh, R., Ghosh, M., et al.: An intelligent offline handwriting recognition system using evolutionary neural learning algorithm and rule based over segmented data points. J. Res. Pract. Inf. Technol. **37**(1), 73 (2005)
4. Nguyen, V., Blumenstein, M.: Techniques for static handwriting trajectory recovery: a survey. In: Proceedings of the 9th IAPR International Workshop on Document Analysis Systems, pp. 463–470. ACM (2010)
5. LeCun, Y., Bengio, Y., et al.: Convolutional networks for images, speech, and time series. Handb. Brain Theory Neural Netw. **3361**(10), 1995 (1995)
6. Bouchain, D.: Character recognition using convolutional neural networks. In: Institute for Neural Information Processing, vol. 2007 (2006)
7. Lauer, F., Suen, C.Y., Bloch, G.: A trainable feature extractor for handwritten digit recognition. Pattern Recogn. **40**(6), 1816–1824 (2007)
8. Gonzalez, R.C., Woods, R.E., Eddins, S.L., et al.: Digital Image Processing Using MATLAB, vol. 624. Pearson-Prentice-Hall, Upper Saddle River (2004)
9. Hu, M.-K.: Visual pattern recognition by moment invariants. IRE Trans. Inf. Theory **8**(2), 179–187 (1962)

10. Cuevas, E., Zaldívar, D., Pérez, M.: Procesamiento digital de imágenes con MAT-LAB & Simulink. Ra-Ma (2016)
11. Hossain, M.Z., Amin, M.A., Yan, H.: Rapid feature extraction for optical character recognition. arXiv preprint arXiv:1206.0238 (2012)
12. Perez, J.-C., Vidal, E., Sanchez, L.: Simple and effective feature extraction for optical character recognition. In: Selected Papers From the 5th Spanish Symposium on Pattern Recognition and Images Analysis: Advances in Pattern Recognition and Applications, Valencia, Spain (1994)
13. Bokser, M.: Omnidocument technologies. Proc. IEEE **80**(7), 1066–1078 (1992)
14. Trier, O.D., Jain, A.K., Taxt, T., et al.: Feature extraction methods for character recognition-a survey. Pattern Recogn. **29**(4), 641–662 (1996)
15. Krizhevsky, A., Sutskever, I., Hinton, G.E.: ImageNet classification with deep convolutional neural networks. In: Advances in Neural Information Processing Systems, pp. 1097–1105 (2012)
16. Simonyan, K., Zisserman, A.: Very deep convolutional networks for large-scale image recognition. arXiv preprint arXiv:1409.1556 (2014)
17. Manoj, A., Borate, P., Jain, P., Sanas, V., Pashte, R.: Offline handwriting recognition system using convolutional network (2016)
18. Otsu, N.: A threshold selection method from gray-level histograms. IEEE Trans. Syst. Man Cybern. **9**(1), 62–66 (1979)
19. de Campos, T.: The Chars74K dataset: character recognition in natural images. University of Surrey, Guildford, Surrey, UK (2012)
20. LeCun, Y.: The MNIST database of handwritten digits (1998). http://yann.lecun.com/exdb/mnist/

Visual Attention Mechanisms Revisited

Cristina Mendoza[1]([⊠]), Pilar Bachiller[1], Antonio Bandera[2], and Pablo Bustos[1]

[1] RoboLab, Universidad de Extremadura,
Avda de la Universidad sn, 10003 Cáceres, Spain
cmendozat@alumnos.unex.es,
{pilarb,pbustos}@unex.es
[2] Universidad de Málaga, Campus de Teatinos, Málaga, Spain
ajbandera@uma.es,
https://robolab.unex.es,
http://www.grupoisis.uma.es

Abstract. Currently robots are evolving into an increasing complexity and have to support a large workload that directly affects their functions. To compensate this situation they must make a better use of their available resources while behaving in a reliable way. The goal of this project is to endow Shelly, the social robot created by RoboLab with a predictive system of visual attention that allows it to maintain an updated internal representation of its environment, providing it with a basic sense of awareness. This improvement allows the robot to foresee simple facts, react to unpredicted situations and integrate changes of the environment in its internal memory. To achieve this level of functionality we have combined overt and covert head movements with an updatable internal model of the environment through a predictive and dynamic attention loop. The system has been developed using the RoboComp framework [21] and the new components have been integrated in the CORTEX cognitive architecture. The implementation is available for public use.

Keywords: Visual attention · Predictive system · Spatial position

1 Introduction

Shelly, the social robot created at RoboLab, is able to interpret voice missions and compute a plan from a domain model, recognize and locate objects [24], safely move through space following a computed path and interpreting people movements and intentions [34], change its trajectory if it finds an obstacle in its path [22], re-compute its path if it gets stuck [33], locate and follow people, maintain simple dialogues and manipulate objects [23]. The complex orchestration of these abilities is possible with the CORTEX cognitive architecture [6], in which a hybrid working memory is shared and updated by a set of agents. Despite its many capabilities, Shelly is not able to react to changes that occur in the environment that involve objects outside the current scope of its plan, to

R. Fuentetaja Pizán et al. (Eds.): WAF 2018, AISC 855, pp. 100–114, 2019.
https://doi.org/10.1007/978-3-319-99885-5_8

integrate new objects in its memory in order to use them in subsequent tasks, or to be aware of its existence and the existence of the rest of objects with respect to itself. This situation is becoming a serious drawback that limits our current efforts with CORTEX. At this point, it seems obvious that what is needed is an attentional system to keep the robot informationally attached to its changing environment, and that is the purpose of this paper. Our research background starts with works published several years ago during the raise of active vision, robotics heads and ocular movements. See for example [1, 28]. However, research on these topics has decreased in recent years partially due to recent advances in RGBD and 360 ° cameras, but also to the recent upswing of deep learning techniques. These new sensing devices almost eliminates the need for stereo vision in indoor environments and simplifies the mechanisms of robotic heads. Also, deep neural networks provide real time and robust labelling of common objects in the scene solving a long lasting problem in Computer Vision. Despite these advances make now possible a robust initial labelling of the scene, a crucial perception problem still remains: how can the robot be aware of its surrounding space and of itself in it? and how can it thoughtfully react to changes that dynamically occur around it? One possible answer that we explore in this paper is that attentional mechanisms coupled with an internal representation of the environment and of the robot, can create a stabilizing dynamics that forever binds the robot to the world [7, 27].

Let's start our proposal by defining awareness. By *being aware* we mean a dynamic state of the robot created by a continuous relationship between it and its environment that is created by:

- A constructive process that incrementally builds up an internal representation of the world.
- A predictive process that computes what will be perceived, overt or covertly, in the immediate future.
- A synchronizing process that minimizes in time the prediction error to keep the internal representation aligned with the world.
- A behaviour generation process that selects what action to do next in order to maximize the utility of the perceived information.

Being aware is thus a dynamic state of the robot in which a near future is continuously re-created by a projection operation on the internal representation. From the past and the present, this operation creates those perceptions that will be most probably perceived in the immediate future. Eventually, when reality kicks in, both are compared and a decision is taken in order to satisfy the demands of the current high level goal of the robot, with the local need to cancel the prediction error. The combination of both streams of actions can be easily achieved if the one with lowest frequency, i.e., the one coming from the high level goal, modulates the one with the highest frequency. When we walk, for example, we generate high frequency eye movements to update our internal representation and keep as safe, while looking once in a while at objects that mark our path to the target.

This approach to awareness is an instantiation of a more general schema that presents intelligence as a prediction-perception-updating machine, and is now getting greater momentum as shown by several recent works in the Philosophy of Mind [8,9]. An interesting variant of this schema that is relevant to this work could be re-stated as prediction-perception-control, and has actually been published several times in the Robotics literature as an effective means to improve the robot's response to changing and unpredictable environments. An classic example of this scheme is the Dynamic Window algorithm [12], which is used to anticipate future trajectories of the robot given different choices of steer and speed, and choose the one that optimizes a cost function related to the target. More recently, the adaptive control system of MIT's four-legged robot Cheeta 3 computes in real time a set of possible foot tip pressures to be sensed after contact with the ground, and the action to be taken after each one. This strategy allows the robot to immediately react to the ground since the response was already computed, skipping the unavoidable delays [2]. In both cases the sensation for the external observer is similar, the robot is aware of its environment since it quickly reacts to unforeseen changes and takes the right choice to keep going towards its goal.

Our hypothesis in this paper is that an implementation of this theory on a robot as a low-level loop connecting perception, action and the internal model (i.e. a highly organized working memory), will provide the robot with a basic sense of being in the world, aware of changes that are detected and internalized. As a result, the CORTEX architecture will benefit from this new functionality and more complex robot missions will be in our reach.

The rest of the paper describes this initial prototype that, working in a restricted environment, shows how the robot Shelly synchronizes its model with its environment and keeps it aligned with it while several external perturbations occur. Section 2 reviews other architectures that include attention and awareness mechanisms. Section 3 describes the prototype created for the robot Shelly. Section 4 discusses the experimental results and Sect. 5 presents some conclusions and future work.

2   Related Work

Even if awareness is a rarely treated topic in the Robotics literature, a key component of it, attention, has received wide interest and many works can be found with interesting implementations. Attention is usually defined as a selection mechanism that reduces the flow of information into the system to those aspects that are useful for the current task [20]. Robots, as well as biological beings, have limited resources to process an enormous amount of perceptual input. Current visual sensors provide more information than can be processed before a choice has to be taken. As the resolution and dynamic range of vision sensors increase, the need for some kind of selection mechanism becomes unavoidable.

Despite the importance of attention, only a few architectures integrate mechanisms for modulating visual data (OMAR [10], iCub's architecture [31], EPIC [16], MIDAS [13] and ARCADIA [4]).

The selection of what visual data to attend can be data-driven (bottom-up) or task-driven (top-down) [18]. The bottom-up attentional mechanisms identify salient regions whose visual features are distinct from the surrounding image features, usually along a combination of dimensions, such as colour channels, edges, motion, etc. Some architectures resort to the classical visual saliency algorithms, such as Guided Search [35] used in ACT-R [26] and Kismet [3], the Itti-Koch-Niebur model [14] used by ARCADIA [4], DAC [25] and iCub [31] or AIM [5] used in STAR [17]. Top-down selection can be applied to further limit the sensory data provided by the bottom-up processing. Many architectures resort to this mechanism to improve search efficiency (ARCADIA [4], DAC [25]). Another option is to use a hard-coded or learned heuristics. For example, MIDAS replicates common eye scan patterns of pilots [13]. The top-down is always looking for new changes at the environment, is an active way to control where itself is and where are other objects. Shelly has a top-down attention system.

First studies of visual attention mechanism conclude that the system of visual attention can be in two states: disengaged or engaged, depending on whether the eyes are moving in saccadic movement or not [11]. Also, shifts of attention can be overt or covert, depending on whether there is a physical movement or a change in the region of interest. Respect to the movements, OMAR, EPIC, MIDAS and ARCADIA do not integrate camera movement, so the attention at these robots is passive, some of them are able to track an object at the scene they are seeing but, if the object disappears from the scene, they are not able to follow it. However, the humanoid robot iCub is able to move eyes and neck but it is a bottom-up system. Shelly must at times maintain attention at a given location, either to wait for an expected new stimulus or to completely process all of a present target's visual information. It must be able to disengage attention in preparation for processing new information, possibly at another location [32].

Another significant item to discuss is object detection and recognition. Hereby, it is important to distinguish several cases: general object detection, object instance detection and object class detection [20]. Object class detection aims at finding all instances of a certain class (any cup). This is a variant of top-down visual search and the one that Shelly implements. To detect objects one can use, for example, SIFT-based object detection [19]. In our case, Shelly uses the YOLO (You Only Look Once) object detector [30], an extremely fast and accurate state-of-the-art, real-time object detection system. The use of YOLO is essential for this work because it allows the detection, with sufficient accuracy, of objects, their location in the image and the class it belongs to.

Summarizing, visual attention is a broad term that encompasses several basic components. These include:

1. Engaging or activating attention in preparation to detect a target of interest.
2. Directing or orienting attention to a specific location in the field.
3. Locking attention on that locus.
4. Suppressing irrelevant information from other locations.

Going from attention to awareness opens many possibilities for discussion. Our interest in this paper is a view of awareness in which attention is coupled with the update of an internal model, and the state of this model modulates the attentional access to the world. This line of thinking suggests that understanding the computational mechanisms underlying active, goal-oriented attention may be a first step toward artificial consciousness [20]. An example of a robot with self-awareness capabilities is ISAC [15]. This robot can achieve similar tasks as Shelly but does not use visual attention mechanisms. To our knowledge, the system we describe bellow is the first to integrate a complex internal representation with an attentional mechanism in a closed loop that minimizes the divergence between what is represented and what is perceived, showings its feasibility in real-world experiments.

3   Proposed Approach

3.1   The Robot Shelly

Shelly (Fig. 1) is the most advanced robot at RoboLab. The purpose of this robot is to assist and extend the autonomy of people with reduced mobility. It has Mecanum wheels that allows it to move in any direction or rotate around itself. Also, it has a LIDAR (Laser Imaging Detection and Ranging) that measures the distance to which nearby obstacles are located; a robotic arm located in the centre for grasping and object manipulation, an ASUS Xtion Pro camera that provides RGBD images, a Kinect 2 camera to locate and follow people around it, 4 NUC mini-computers connected to a switch where the components are executed, a touch screen, batteries and power and recharge electronics.

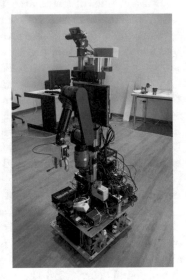

Fig. 1. The robot Shelly.

Shelly executes ocular saccadic movements. A saccade is a rapid movement of the eye from one position to another. In saccades, the eyes move during a time interval between 20 and 200 ms. In this phase, the eyes do not extract information from the stimulus. At each saccade or abrupt jerking of the eye, an eye fixation follows, in which the eyes remain almost static for 250 ms. The extraction of information takes place in this phase [29].

In our system the robot performs saccades to focus attention from one object to another or to track moving objects. This movement is carried out by a two degrees of freedom robotic eye.

3.2   Inner Model Representation

Attention requires some sort of internal memory to store the places already visited and to decide where to move next, overtly or covertly. Simple maps representing visited positions in ego-space is a frequent solution but they are not expressive enough to reason about the spatial arrangement of the space and the objects in it.

Shelly represents its environment using a hybrid kinematic-symbolic scene graph that relates the known parts of the world and the robot itself, as defined in the CORTEX architecture. This representation can be serialized into an XML format, written and recovered every time a new execution of the whole system starts. This internal representation along with the attention mechanism form the basic system that allows the robot to reach awareness, in the sense defined above. Part of the attention process occurs before information reaches awareness; it is pre-attentive and can use a parallel processing strategy to acquire data. The attentive process is conscious and uses a serial discovery strategy based on selected data from the pre-attentive acquisition process, provided by the YOLO network. From YOLO and the depth information in the RGBD camera, we obtain the type and position of each object in the camera's system of reference. Using the dynamically updated kinematic graph of the robot, the position in the scene is easily computed. To represent and store the objects in the environment, the following attributes are considered:

- *type* of the object
- *name* of the object
- *identifier*
- *bounding box*
- *projection* of the bounding box on the image
- *position* of the object in the world
- *number consecutive frames* without detecting the object

The bounding box is the minimum cube that contains the object. Currently, the system only works with cups and glasses, and it is assumed to be a $80 \times 80 \times 80$ mm cube large enough to completely fit the object.

The kinematic graph is extended with several lists that track temporal changes in local state:

- *listObjects*: complete list of existing objects in the world.
- *listYoloObjects*: list of objects detected by Yolo.
- *listCreate*: list of detected objects not included in the internal representation.
- *listVisible*: list of current visible objects stored in memory.
- *listDelete*: list of visible objects stored in the internal representation of the robot that have not been detected by Yolo.

3.3   Attention Mechanism: Thermal Map

To control the camera movements the attentional system uses thermal maps that represent the need for attention of the different areas of the scene. Specifically,there is a thermal map associated to each container of objects. In the current implementation, only tables are initially considered as object containers, thus, the plane of a table acts as a control surface that drives attention according to the following considerations:

- The map is divided into cells expressed in coordinates of the table plane. Every cell has its own temperature value which will be modified according to the need for attention.
- Each cell of the map has an initial value of temperature of zero, that will increase or decrease depending on the location of the objects on the table at each moment and the position of the camera.
- The map will be continuously cooled with a consistent constant descent temperature for the entire map, except for the areas where an object is located. Cells that touch an object will cool more quickly because they require more attention, since these cells correspond to the most likely change areas.
- The map will heat up depending on the area where the camera is focused using a Gauss function (Fig. 2). This means that, if a cell enters the projection of the camera at a certain time, its temperature will increase according to its position in the image. Thus, if the cell is near the image centre, it will heat up faster than if it were near the image limits.
- The camera moves towards the coldest areas of the map. This guarantees that every area of the table will be eventually explored.

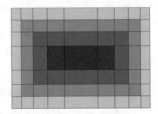

Fig. 2. Kernel for the heating filter

This strategy to drive attention on the surface of containers of objects -tables- could be extended to other objects in the hierarchical representation of the environment while being modulated by the current task of the robot. Thus, every cup could have a thermal map to switch the attention between the body and the handle, in case the robot needs to grasp it. At a higher level in the hierarchy, a thermal map associated to the floor would make the robot explore the room and react to changes in it, so it can safely move around. At every level, cooling and heating dynamics of each thermal map would be defined and conditioned by the task and goals of the robot.

3.4   Design of the Attentional Agent

The visual attention system has been developed as a RoboComp component connected to the other perceptual and motor components taking part in the process. This new component implements a state machine (Fig. 3) that detects changes in the environment and updates the internal representation of the world according to these changes. A state machine is a system behaviour model with inputs and outputs, where the outputs depend, not only on the current input signals, but also on the previous ones.

The states in this machine are:

- *Predict*: initial state that receives a fresh image and a reads the orientation of the camera, in order to compare the real and the synthetic world later on. When finished it goes to the *YoloInit* state.
- *YoloInit*: intermediate state that sends the new image to the Yolo server. It moves to *YoloWait* to wait for the labels.
- *YoloWait*: intermediate state that waits for the Yolo server to finish processing the image. It receives the labels of the different objects in the scene and moves to the *Compare* state.
- *Compare*: compares the objects of the synthetic world with the labels provided by the Yolo server. In doing that it compares the predicted represented world with the real one. It moves to the *Stress* state.
- *Stress*: if no change is detected between the synthetic and real worlds, it does nothing, but if there has been any change in the position or existence of an object, the internal representation is updated with the perceived novelties. It goes to the *Predict* state.
- *Moving*: a special state that can be reached from any other state and waits until the camera motors are stopped. This wait guarantees a new clear image and when done it always jumps to the *Predict* state, where the current camera position is obtained.

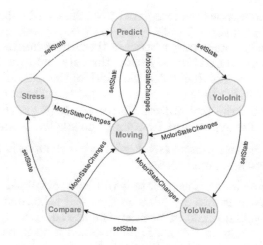

Fig. 3. Component state machine.

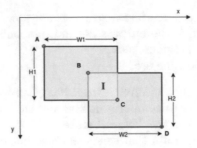

Fig. 4. Intersection diagram.

Comparison Between the Internal Representation and the Real World.
The comparison between the internal representation and the perceived world is
done by trying to associate each perceived object with a synthetic, predicted
one. Thus, for each object in the list of detected objects (*listYoloObjects*), if it
is of a known type, in our case "cup", and it has not yet been assigned to any
of the synthetic objects (memory objects), it is processed as follows:

1. Take the rectangle defining the real object.
2. Compute the rectangle that results from the projection of the bounding box
   of the synthetic object in the image.
3. Compute the intersection (Fig. 4) of the two rectangles and the corresponding
   intersection area (I).

$$B = (max\{x_A, x_B\}, max\{y_A, y_B\})$$
$$C = (min\{x_C, x_D\}, min\{y_C, y_D\})$$
(1)

The intersection area is then calculated from the coordinates of the intersect-
ing rectangle, B and C.

4. Obtain the displacement vector between the two rectangles.
5. If the area of the intersection is greater than zero and the length of the displacement vector is less than twice the width of the rectangle of the synthetic
   object, a new candidate is created and added to a list of candidate objects.
   This list contains the real objects that potentially match a synthetic one.
   Objects in this list are ordered from larger to smaller intersection area.

After completing this process (see Fig. 5), what remains is the application of
the update-remove-insert operations:

- if a synthetic object has a candidate it will be a real object that is very close
  to its position, so the first candidate in the list is taken and the synthetic one
  is marked as explained and its position is updated.
- if there is any synthetic object left unexplained it must be erased. For some
  reason it has disappeared from the field of view.
- if there is any object in the list of detected objects that does not correspond
  to any synthetic object, it means that there is an object that the system did
  not contemplate, but it has appeared in the scene. Then, a new object is
  created and added to the list of new objects (*listCreate*).

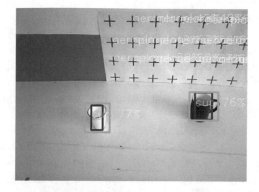

Fig. 5. Matching between synthetic (green) and detected (blue) objects.

4 Experimental Results

To test the proposed attentional control system a series of experiments have been conducted within *RoboLab*'s Autonomy Lab. This scenario includes two tables placed forming an "L", as is shown in Fig. 6. The robot head (camera and neck) has been mounted on a tripod since, for these initial experiments, it is not necessary to use the complete robot base. For each experiment a video has been recorded showing the robot's behaviour from three different views. Figure 6 shows an example snapshot. In this figure three different parts can be distinguished: the real environment (left), the internal representation of the environment (top-right) and the view from the robot camera (bottom-right).

Fig. 6. Snapshot of one of the recorded videos.

The aim of the first experiment is to test the ability of the robot to track objects. The expected behaviour is for the robot to keep the attention fixed on an object while it is moving. This implies that the robot has to continuously recalculate the position of its motors to maintain the camera centered on the object. The thermal map plays an important role in this behaviour since moving objects intensely cool a new area of the table, causing the camera to move and fixate on this new attention point. If the object ceases its movement the robot holds the attention for a few seconds, but then it diverts it to another colder area of the table. To perform this test we have used a single cup that has been moved on the table with the help of a person. The video of this test is available in the following link: https://youtu.be/TZZbTtW21vE.

The second test consists of alternating attention between two cups. To do this, two cups are initially placed on a table in the internal representation. The robot begins observing its environment, stores the position of the first cup that it finds and fixates on it. Since the cup is not moving the robot can divert its attention to other colder zone. As there are two preloaded cups, the map will cool the corresponding table areas more intensively until it checks that they do not exist in the real world. When the robot stops attending to the first cup that has been found, its attention is directed to the new coldest zone which is the area of the synthetic cup in the internal representation. If it finds a real cup on the way, it focuses on it until its temperature increases enough to divert its attention again. Then, it goes back to find the first cup that remains in the same position. This behaviour continues for a while alternating between both cups until other areas of the table demand attention, i.e., present a very low temperature in the thermal map. Then, the camera focuses in a new selected area which corresponds to the coldest cell in the table map. Since there is no other object in this third area, it heats up quickly and the previous cups captures the attention again. If, after looking at this third zone, a third object was found, the attention would alternate among the three existing objects. Results from this test are available through the following link: https://youtu.be/bkswvoK188M.

For the third experiment the previous two tests have been combined: maintenance of attention and *tracking* with more than one object (see the Fig. 7). The experiment starts with the robot tracking a moving cup. At a certain moment, the cup stops moving and a new cup appears in the scene. Since this cup has not been seen before it obtains the focus of attention. Then this second cup starts moving. Although both cups are still in the scene, the one that is moving is the one that captures the attention since it varies the zone of minimum temperature. When the second cup stops moving, attention returns to the first cup. If the first cup starts moving again, it keeps the robot's attention until it stops. We stop one cup next to the other. This implies that both must receive the same degree of attention because both of them are static. Then, the attention starts alternating. A video with the results of this experiment can be found at the following link: https://youtu.be/DoxEQk_JEi0.

To complete the experiments we have tested the robot's behaviour in situations in which the attention must be divided among different tables. The robot can pay attention to more than one table, but only to one at a specific time. This means that the focus is on a single table, even if the system captures more information and keeps updating its environment. Changing the focus of attention from one table to another is produced, at a higher control level, when the minimum global temperature goes under a certain threshold. When this happens there is another table that requires more attention. Afterwards, the system analyzes the map of the new table and looks for the coldest area to center the attention on it. In this experiment, when the camera reaches the second table, it is empty. Then, a cup is added and it gets the focus of attention. The attended table is maintained for a while until the first table again requires attention. The attention time on each table can be adjusted and is related to the cooling rates. In the following link, an experiment with a small time interval between table changes can be found: https://youtu.be/kT-OAfUe6FA.

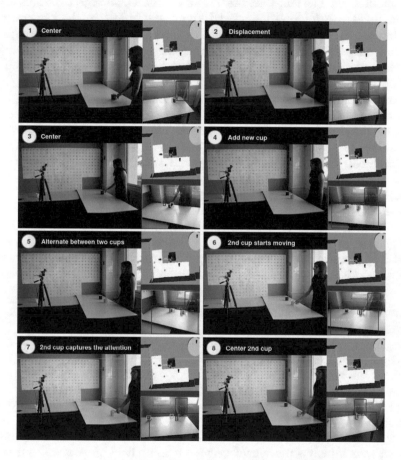

Fig. 7. Sequence of snapshots of one of the recorded videos.

5   Conclusions and Future Work

This paper presents a proposal for an awareness subsystem to be integrated in the CORTEX cognitive architecture. The algorithms described above create a sustained dynamic loop coupling what is perceived in the world and what is represented inside the robot. This dynamics minimizes the differences between both by applying simple editing operations such as insertion, deletion and update at the object level. A crucial insight of this work is the realization that prediction and correction at the object level, i.e. comparison of two lists of labels and coordinates, is much more efficient than working at the sensor level, i.e creating synthetic images from object change predictions, and much less prone to alignment errors. This feature would not had been possible without a system like YOLO, that shortcuts to milliseconds what a few years ago was a very long and unreliable computation. The fact that the output of YOLO, an object label, is already a direct handle to information stored about the object in the system, opens many interesting possibilities for future research. For example, the comparison between synthetic and perceived objects could be constrained by semantically rich information about the current context, the task or the functional properties of the objects involved.

In the field of social robotics, where CORTEX is being currently used, awareness is becoming an urgent need. Social awareness relates to being aware of your environment, what is around you, as well as being able to accurately interpret the emotions of people with whom you interact. Although that is a much more ambitious goal, the results presented here are a step in this direction, providing an initial infrastructure from which to test and validate new hypothesis.

The system as it is now must be improved to be faster and more sensitive, but it is an interesting first approach to a much more complex model of attention. A more efficient execution cycle can be achieved by parallelization of certain parts, and a shorter period between frame captures when the eye is saccading can be achieved by a more sophisticated velocity control that predicts the end of the movements. Along with these general improvements, additional ongoing work is being considered:

- Extend the thermal map control to the complete hierarchy of world objects represented in the inner model of the robot.
- Integrate the head with the robot's base movements to extend its awareness domain to complete rooms.
- Extend the types of objects detected by the system. This should be possible within the limits of YOLO, since it is the employed object detector.
- Add a "velocity" property to the visible objects in order to improve the tracking and prediction process.
- Add the smooth pursuit movement to the camera so it is able to track an object in continuous movement keeping it in its field of view without sudden jumps. This mode would increase the frame rate and, therefore, the sensitivity of the system.
- Replace the current heat map algorithms with partial differential equations to obtain a smoother behavior.
- Add robustness to the insertion/deletion operations based on learned time or repetition thresholds, i.e. an object must appear or disappear in several consecutive frames to be added or deleted. Also other criteria based on semantic information could be applied.
- Explore ways of creating and learning new unseen objects from the perceived shape.

An advanced model of this system would observe the real environment and dynamically create its own internal model, based on a combination of measuring, prediction, correction and conceptualization. We believe that this initial work is a step in the right direction.

References

1. Bachiller, P., Bustos, P., Manso, L.J.: Attentional selection for action in mobile robots. In: Advances in Robotics, Automation and Control, pp. 111–136 (2008)
2. Bledt, G., Wensing, P., Sangbae, K.: Policy-regularized model predictive control to stabilize diverse quadrupedal gaits for the MIT cheetah. In: IEEE International Conference on Intelligent Robots and Systems, Vancouver, BC, Canada, pp. 4102–4109 (2017)
3. Breazeal, C., Scassellati, B.: A context-dependent attention system for a social robot. In: In IJCAI International Joint Conference on Artificial Intelligence, San Francisco, CA, USA, vol. 2, pp. 1146–1151 (1999)
4. Bridewell, W., Bello, P.F.: Incremental object perception in an attention-driven cognitive architecture. In: Proceedings of the 37th Annual Meeting of the Cognitive Science Society, Atlanta, Georgia, pp. 279–284 (2015)
5. Bruce, N., Tsotsos, J.: Attention based on information maximization. J. Vis. **7**, 950–952 (2007)
6. Calderita, L.V.: Deep state representation: an unified internal representation for the robotics cognitive architecture cortex. Master's thesis, University of Extremadura, Cáceres, Spain (2016)
7. Carpenter, R.H.S.: Movements of the Eyes, 2nd edn. Pion Limited, London (1988)
8. Clark, A.: Surfing Uncertainty. Oxford University Press, England (2016)
9. Danks, D.: Unifying the Mind. MIT Press, Massachusetts (2014)
10. Deutsch, S.E., Macmillan, J., Camer, M.L., Chopra, S.: Operability model architecture: Demonstration final report. Technical Report AL/HR-TR-1996-0161 (1997)
11. Fischer, B., Breitmeyer, B.: Mechanisms of visual attention revealed by saccadic eye movement. Pergamon Journals Ltd (1987)
12. Fox, D., Burgard, W., Thrun, S.: The dynamic window approach to collision avoidance. Robot. Autom. Mag. **4** (1997)
13. Gore, B.F., Hooey, B.L., Wickens, C.D., Scott-Nash, S.: A computational implementation of a human attention guiding mechanism in MIDAS v5. In: International Conference on Digital Human Modelling, California, USA (2009)
14. Itti, L., Koch, C., Niebur, E.: A model of saliency-based visual attention for rapid scene analysis. IEEE Trans. Pattern Anal. Mach. Intell. **20**, 1254–1259 (1998)
15. Kawamura, K., Dodd, W., Ratanaswasd, P., Gutiérrez, R.A.: Development of a robot with a sense of self. In: IEEE International Symposium on Computational Intelligence in Robotics and Automation, Espoo, Finland (2005)
16. Kieras, D.E., Wakefield, G.H., Thompson, E.R., Iyer, N., Simpson, B.D.: Modeling two-channel speech processing with the epic cognitive architecture. Top. Cognit. Sci. **8**, 291–304 (2016)
17. Kotseruba, I.: Visual attention in dynamic environments and its application to playing online games. Master's thesis, York University, Toronto, Canada (2016)
18. Kotseruba, I., Tsotsos, J.K.: A review of 40 years in cognitive architecture research core cognitive abilities and practical applications. Cornell University Library (2018)

19. Lowe, D.G.: Distinctive image features from scale-invariant keypoints. Int. J. Comput. Vis. (IJCV) **60**, 91–110 (2004)
20. Mancas, M., Ferrera, V.P., Riche, N., Taylor, J.G.: From Human Attention to Computational Attention: A Multidisciplinary Approach, 1st edn. Springer, Crete (2015)
21. Manso, L., Bachiller, P., Bustos, P., Núñez, P., Cintas, R., Calderita, L.: RoboComp: a tool-based robotics framework. In: SIMPAR. LNCS, vol. 6472, pp. 251–262. Springer (2010)
22. Manso, L.J., Bustos, P., Bachiller, P.: Multi-cue visual obstacle detection for mobile robots. J. Phys. Agents **4**, 3–10 (2010)
23. Manso, L.J., Bustos, P., Bachiller, P., Franco, J.: Indoor scene perception for object detection and manipulation. In: 5th International Conference Symposium on Spatial Cognition in Robotics, Rome, Italy (2012)
24. Manso, L.J., Gutiérrez, M., Bustos, P., Bachiller, P.: Integrating planning perception and action for informed object search. Cognit. Process. **19**, 285–296 (2018)
25. Mathews, Z., Bermudez I Badia, S., Verschur, P.: PASAR: an integrated model of prediction, anticipation, sensation, attention and response for artificial sensorimotor systems. Inf. Sci. **186**, 1–19 (2012)
26. Nyamsuren, E., Taatgen, N.A.: Pre-attentive and attentive vision module. Cognit. Syst. Res. 211–216 (2013)
27. Pahlavan, K.: Active Robot Vision and Primary Ocular Processes, 1st edn. Royal Institute of Technology Stockholm, Computational Vision and Active Perception Laboratory (CVAP), Sweden (1993)
28. Palomino, A., Marfil, R., Bandera, J.P., Bandera, A.J.: A novel biologically inspired attention mechanism for a social robot. EURASIP J. Adv. Sig. Process. 1–10 (2011)
29. Purves, D., Augustine, G., Fitzpatrick, D., Hall, W., Lamantia, A., Mcnamara, J., Williams, S.: Neuroscience, 3rd edn. Sinauer Associates (2004)
30. Redmon, J.: Yolo: Real-time object detection (2018). https://pjreddie.com/darknet/yolo/
31. Ruesch, J., Lopes, M., Bernardino, A., Hornstein, J., Santos-Victor, J., Pfeifer, R.: Multimodal saliency-based bottom-up attention a framework for the humanoid robot iCub. In: Proceedings of the IEEE International Conference on Robotics and Automation, Pasadena, CA, USA, pp. 962–967 (2008)
32. Steinman, S., Steinman, B.: Topics in Biomedical Engineering International Book Series, Models of the Visual System. Springer, Boston (2002)
33. Um, D., Gutiérrez, M.A., Bustos, P., Kang, S.: Simultaneous planning and mapping (SPAM) for a manipulator by best next move in unknown environments, Tokyo, Japan, pp. 5273–5278. IEEE (2013)
34. Vega, A., Manso, L.J., Macharet, D.G., Bustos, P., Núñez, P.: A new strategy based on an adaptive spatial density function for social robot navigation in human-populated environments. In: REACTS Workshop at the International Conference on Computer Analysis and Patterns, CAIP. Ystad Saltsjbad (2017)
35. Wolfe, J.M.: Guided search 2.0 a revised model of visual search. Psychon. Bull. Rev. **1**, 202–238 (1994)

Change Detection Tool Based on GSV to Help DNNs Training

Carlos Gómez Huélamo[1(✉)], Pablo F. Alcantarilla[2], Luis Miguel Bergasa[1], and Elena López-Guillén[1]

[1] Department of Electronics, University of Alcala, Alcalá de Henares, Spain
carlos.gomezh@edu.uah.es, {luism.bergasa,elena.lopezg}@uah.es
[2] SLAMcore Ltd., London, UK
pablofdezalc@gmail.com
http://www.robesafe.es/personal/bergasa/
http://www.robesafe.es/personal/pablo.alcantarilla/

Abstract. We present a system to carry out the automatic detection of structural changes through a Deconvolutional Neural Network (DNN) in images synthesized from panoramas provided by an online and open source map tool, Google Street View (GSV). Our approach is motivated by the need of more efficient and frequent updates on large-scale maps for autonomous driving applications. To train and evaluate our DNN we build a geolocation database, an order of magnitude larger than other existing datasets, based on pairs of images and their corresponding ground truth that shows changes detection over time. A tool has been implemented to guide manual annotation of changes using panoramas all over the world. The tool chains the panoramas and depth maps creation, the image synthesis and the labelling synthesized images generating their groundtruth. Finally, a DNN has been trained to automatically detect changes validating our methodology by using the obtained dataset, yielding better results that other state-of-the-art approaches.

Keywords: Google Street View · Change detection · DNN
Keypoint · Training/validation instances · Matching

1 Introduction

When viewed at the scale of cities and over periods spanning seasons or years, we realized that the urban visual landscape is a highly dynamic environment, with many navigational signs such as buildings, traffic signs and other structures located at both sides of the road being highly added or removed [1,2]. From the point of view of an autonomous driving system, keeping an updated map of certain places of interest is essential. The higher the frequency with which maps are updated, the more robust the navigation system will be.

In this work we address the problem of efficient maintenance of a map by detecting structural changes using a powerful tool made available to the public,

© Springer Nature Switzerland AG 2019
R. Fuentetaja Pizán et al. (Eds.): WAF 2018, AISC 855, pp. 115–131, 2019.
https://doi.org/10.1007/978-3-319-99885-5_9

online and totally free such as Google Street View (GSV). The greatest advantage of using GSV is that structural changes over the years all over the world can be recovered by using the different laps of the Google Car in the same places. This information can be very interesting for historical, engineering, architectural (cadastres) or sociological analysis, among others. According to driving applications, keeping maps up-to-date is not only useful for robust navigation, but can also provide benefits in monitoring the availability of parking spaces or road closures and deviations due to road works. Detecting structural changes in urban environments from GSV images is a challenging problem, as illustrated in Fig. 1.

Fig. 1. Examples on images that present challenging changes generated from Google Street View, taken from our dataset. Left and right columns show a pair of images (older and newer ones), both being as aligned as possible.

Note that all examples in Fig. 1 assume changes in luminosity, weather and station, and what makes it even more challenging is the point of view.

Images taken at different times show a great variability that can be induced by structural changes (construction/demolition of buildings/traffic signals, among others), but also due to nuisances generated by changes in point-of-view (the main problem in this work), environmental conditions (luminosity, weather or season) or dynamic changes (pedestrians, vehicles or vegetation). The main task of this work is the detection of structural changes. Therefore, in order to successfully distinguish among noises and real structural changes, the detection method must be able to model both, not-detecting the first ones and detecting the second ones.

So far, structural changes are manually detected on images, being a very time consuming task. For this purpose we present a real powerful tool able to process geolocations all over the world, getting as output pairs of images and its differential groundtruth (structural changes between pairs of aligned images) in order to serve as training/validation dataset for neural network implementations.

Based on the current success of Convolutional Neural Networks (CNNs) in different image processing task: image classification [3], semantic segmentation [4] and site recognition [5], our tool has been validated by using a deconvolutional deep architecture [6] in order to detect structural changes. This DNN, shown in Fig. 2, takes as inputs a pair of aligned images of the same geolocation and returns a highlight of the structural changes present between both based on pixel-wise classification.

The main contributions of our paper are the following:

1. We propose a powerful tool that combines an on-line and open source map tool such as GSV and a self-code able to process geolocation databases, obtaining a dataset useful to train a Deconvolutional Neural Network in the task of detecting structural changes.
2. We provide a new dataset for the task of detecting urban changes that presents an order of magnitude larger than other existing datasets. In addition, it contains challenging landscapes, due to variations of light, weather or seasons.
3. We validate our tool and dataset by using an improved deep deconvolutional architecture compared to previous works [7] that significantly improves the results of change detection with hand-crafted descriptors [8–10] and a CNN-based approach [11] on the street view change detection task.

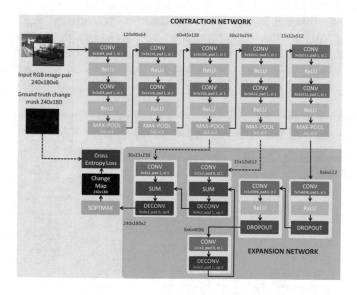

Fig. 2. DNN architecture adopted throughout this work.

2   Change Detection Tool Architecture

Image-based change detection is a major problem in robotics and computer vision. This detection can increase the efficiency of maintaining a 3D map by updating only the areas of change [12] or allowing a system to learn about the nature of objects in the environment by segmenting while they are changing [13].

In order to train the DNN, our tool must generate training/validation instances, consisting in two aligned RGB images of the same place in different time and a groundtruth image with the structural differences between the two images. Flowchart Fig. 3 shows the architecture followed in our tool. As inputs, it takes two different in time links for the same geolocation. As output, it returns training packages constituted by pair of aligned images (from where the DNN will highlight the main structural changes perceived) and its differential groundtruth in order to validate the results.

2.1   Creating the URL Database

The first step is to obtain two images for the same geolocation at different times. The conditions that should have the geolocation (both situations included) must take into account these requirements: Low or no presence of pedestrians, vehicles and vegetation, little difference in brightness and moderate structural changes in order to carry out the subsequent extraction of keypoints and alignment. These structural changes should not reach the stage where is impossible to realize if both images correspond to the same information. Figure 4 shows the GSV interface with the option of the time machine framed in a red square. The user can obtain two different in time links for the same geolocation (eg: March 2015 and February 2017). After that, these links are stored in a geolocation database linked to our application. For that reason, if an user modify the database, the tool automatically will detect the change and operate in an updated way.

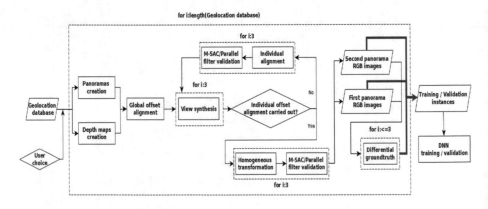

Fig. 3. Flowchart adopted in our tool.

Notice that the more effective the search of geolocations (better coincidence in position and camera orientation), the more coincident will be the information represented in each pair of images, so the easier and more useful will be these images in the training process.

Fig. 4. GSV interface. Framed in a red box, represents the time machine that allows the user to navigate through different for the same geolocation.

2.2 Creating the Panoramic Image

Both panoramic image generation and its depth map are based on the three-dimensional reconstruction of cities from GSV [14]. GSV does not provide the user with a panoramic image, but thanks to these previous geolocation links we must perform a panorama reconstruction by using an API provided by GSV:

https://geo0.ggpht.com/cbk?cb_client=maps_sv.tactile& authuser=0 & hl = en& panoid=%s&output=tile&x=%d&y=%d&zoom=3

The arguments to be introduced to the API are the panorama identifier (or panoid) and the horizontal (x) and vertical (y) position of the specific tile we are analyzing. By default, GSV divides the panorama into 21 tiles, grouped into 7 columns and 3 rows. As illustrated in Algorithm 1, introducing both panoid and a row and column identifier (ix, iy), rows and columns are traversed and each tile analyzed for a given geolocation. Notice that when representing the information of each tile into the panorama, the function Ipan(ix, iy, :) = im means that this portion of the panorama adopts the whole tile information (512 x 512 pixels) including the RGB channels. Finally, panorama dimensions are 1536 x 3584 pixels. Figure 5 shows an example of estimated panorama reconstruction, where the panoramic projection of the spherical image caught by GSV car be perfectly seen.

2.3 Computing the Depth Map

Now that we have the panoramic image, we need to retrieve its corresponding depth map. Google Maps REST API lets us to download a depth map JSON representation from the following link:

https://maps.google.com/cbk?output=json&cb_client= maps_sv & v = 4 & dm = 1&pm=1&ph=1&hl=en&panoid=%s(3)

If we decompress the downloaded file it can be observed an alphanumeric file that contains the distance from the camera to the nearest surface for each pixel of the panoramic image. After decoding from Base64 data, transforming into UINT

Algorithm 1. Computing the panorama in function of chosen resolution

Input: Panoid
Output: Estimated panorama reconstruction

1: for ix = 0 $\longrightarrow$ xtiles - 1
2: for iy = 0 $\longrightarrow$ ytiles - 1
3: $st\_url = sprintf('https : //geo0.ggpht.com/cbk?cb\_client = maps\_sv.tactile\&authuser = 0\&hl = en\&panoid = \%s\&output = tile\&x = \%d\&y = \%d\&zoom = 3', panoid, ix, iy);$
4: im = imread(st_url);
5: Tile composition:
$$ypos = iy \cdot tile\_height + 1 \longrightarrow (iy + 1) \cdot tile\_height;$$
$$xpos = ix \cdot tile\_width + 1- \longrightarrow (ix + 1) \cdot tile\_width;$$
6: Ipan = (xpos,ypos,:) = im;

8-bits array and then parsing the raw data in order to obtain the header information (that contains the number of planes, width and height of the decompressed depth map). This header information is pretty useful since parsing again its values we realize that each pixel of this 512 x 256 grid corresponds to one of these planes (adjacent pixels with the same depth are contained in the same plane), each plane featured by its normal vector and minimum distance to the camera, respectively plane.n and plane.d in Algorithm 2. As mentioned above, the panorama is reconstructed by using spherical image caught by GSV. By European agreement, considering spherical coordinates, ϕ angle or azimuthal angle goes from 0 to 360 degrees (0 to 2π radians). On the other hand, θ angle or colatitude angle goes from 0 to 180° (0 to π radians). Notice that an offset of $\frac{\pi}{2}$ is added to azimuthal angle in order to obtain a more realistic representation. The variable ur in Algorithm 2 represents the vector that joins a given pixel with the center of the camera, that is, the vector to transform our spherical coordinates into cartesian coordinates. To determine the depth of each pixel, it must be calculated the intersection of a ray that starts at the center camera [14] (whose vector is ur) and its corresponding plane. It must be taken into account that if planeIndex is equal to 0, the depth associated to that pixel will be maxDepth (input parameter). Otherwise, if planeIndex is higher than 0, the associated depth will be the absolute distance from the center of the camera to that pixel, represented by pixdistance, as illustrated in Algorithm 2. Iterating for all planes (all pixels per plane), we can compute our depth map as a bidimensional array of 512 x 256 elements each one 32 bits length. Figure 5 shows an example of depth map generation based on panoramic projection of the spheric image caught by GSV.

2.4   View Synthesis

After elaborating the panoramic image and associated depth map a binary differential groundtruth could be generated in order to train the DNN [15]. However, previously images must be aligned. To do that we synthesize images from the panorama, using the panoramic image and its depth map (Fig. 5), because is eas-

---

Algorithm 2. Computing the depth map

Input: Header, planes
Output: Estimated depth map

1: depth = zeros $(header \cdot height, header \cdot width)$
2: for h = 0 $\longrightarrow$ header.height - 1
3: for w = 0 $\longrightarrow$ header.width - 1
4: planeId $\longleftarrow$ $index\left[h \cdot header.width + w + 1\right]$;
5: $\phi \longleftarrow 2\pi \cdot \frac{header.width - w - 1}{header.width - 1} + \frac{\pi}{2}$;
6: $\theta \longleftarrow \pi \cdot \frac{header.height - h - 1}{header.height - 1}$;
7: $ur \longleftarrow [sin(\theta) \cdot cos(\phi), sin(\theta) \cdot sin(\phi), cos(\phi)]$
8: If planeId > 0 then

$$plane \longleftarrow planes\,[planeId];$$
$$t = \frac{plane \cdot d}{ur \cdot plane.n};$$
$$pix\_camera\_vector \longleftarrow v \cdot t;$$
$$pix\_camera\_distance \longleftarrow \sqrt{pix\_camera\_vector \cdot pix\_camera\_vector'};$$
$$depthMap\,[y \cdot w + (w - x - 1)] \longleftarrow pix\_camera\_distance;$$

9: else

$$depthMap\,[y \cdot w + (w - x - 1)] \longleftarrow maxDepth;$$

---

ier to implement structural changes in smaller images and geolocation efficiency is a critical point.

This synthesis pretends to deal with the problem of visual shift and visual recognition of the place at large-scale where the scene suffers important changes, such as illumination, wear through time or explicit structural changes.

Planar structure of the depth map provides a really coarse 3D structure of the scene that will be represented in future synthetic views. Nevertheless, its precision is good enough for our purposes. The main objective of synthesizing both panoramas is to obtain pairs of images with approximately the same field of view so as to appreciate possible changes.

The flowchart adopted by our tool, shown in Fig. 3, defines a feedback process in relation with the view synthesis process. First, an initial synthesis of the first panorama is carried out. Second, a global offset alignment is calculated and applied to the second panorama with the purpose of this second panorama with the same point-of-view of the first one. An individual extra offset is calculated, if required, based on matching techniques from each pair of images (from first and second panorama) in order to align them horizontally as best as possible. The synthesis process is explained in detail in paper [15]. For our purposes, a 5° pitch angle and a field of view of 35° are considered for our synthetic images from the center of the panorama. Notice that really these synthetic images do not exist but there are created by using GSV information. This is one of the highest contributions of this paper because our tool is able to generate different point-of-view images in an easy way. With the purpose of aligning both panoramas with a global offset, the tool is able to return the first panorama initial column from where the view synthesis is going to start, that is, zero degrees in our yaw vector, that represents the array that stores the view synthesis initial angles in spherical projection. In order to obtain a precise global offset alignment the tool stores the polyline manually drawn by the user, which must correspond to the same information that showed in the first initial column. An example can be observed in Fig. 6. Global alignment is based in the following equation:

$$Global\_offset = \frac{mean(User\_line(:,1)) - mean(in\_col)}{ImWidth} \cdot 360$$

Fig. 5. Inputs for view synthesis. Left side, generated panoramic image. Right side, associated depth map.

As mentioned, ImWidth is equal to the panorama width, in this case 3584 pixels. This equation represents the equivalence in degrees between the horizontal offset presented in both panoramas and the projection of the panorama over a horizontal plane. If the global offset was negative (red column more to the right than blue column), it must be performed a positive anti-offset (adding the corresponding degrees to the heading angle of the camera). Otherwise, a negative anti-offset is carried out. Pitch angle and field of view are the same in both cases.

Fig. 6. Global alignment interface. It can be seen in red the first panorama initial column, calculated by the tool, and the second panorama initial column, drawn by the user.

2.5   Individual Pair Images Alignment

Despite the view synthesis process is generated from globally aligned panoramas, it is not weird finding these camera points-of-view presenting strong displacements in terms of rotation and/or rotation. To ensure a fine alignment, it is convenient to individually align each pair of images by using matching techniques.

The previous global alignment is calculated determining the difference between initial columns. On the other hand, this additional individual pair images offset is calculated determining the arithmetic average among horizontal differences of theoretical good pairs of inliers (Algorithm 3) validated after the M-SAC and parallel filter (Algorithm 4).

With this double offset correction (global and individual offset) these pair of images are well aligned. However, in order to generate in an easier way the differential groundtruth, second panorama synthetic views are generated again, as illustrated in Flowchart Fig. 3, by using a homogeneous transformation to correct errors in terms of rotation or vertical translation. Notice that these homogeneous transformed images will not be introduced into the network but it will be useful in order to generate the differential groundtruth.

To carry out both alignments (individual and homogeneous transformation), some keypoints must be found in both images of each pair of images, matching those that correspond to the same information. Three keypoints detection techniques are applied, such as Harri's corner detector [16], SURF technique (Speed-Up Robust Feature) and SIFT technique (Scale-Invariant Feature Transform) [9]. Depending on three different techniques instead of only one provides us a greater flexibility facing condition changes, such as rotation/translation of the point of view, different illumination, vegetation or season. After obtaining these required keypoints, a M-SAC (M-Estimator SAmple and Consensus) filter is applied to remove the outliers. The main advantage of M-SAC filter versus the classic RANSAC (RANdom SAmple Consensus) technique [17] is that RANSAC technique can be sensible to the chosen threshold that determines which pair of keypoints is valid facing to the whole set of keypoints or not. For that reason, M-SAC variant is used based on evaluating the set of pairs quality calculating its probability. However, after checking some panoramas, we observed that after applying this M-SAC filter, just with two pair of erroneous keypoints outliers, the evaluated individual offset was really different from the real one and even after the second stage of this feedback (homogeneous transformation).

For that reason, we create a parallel filter based on reinforcing outliers discard, based on the Algorithm 4. It can be proved that if the second panorama section is located following the first one (both presenting 640 x 480 pixels, RGB images and groundtruth dimensions) can be considered a vector relating the second panorama section keypoint (end) and its counterpart in the first panorama section (origin). With elementary trigonometry the module and angle of this vector can be calculated.

Considering the module and angle of the initial vector characterized as our reference, a pair of keypoints will be validated if its module and angle is adjusted

---

Algorithm 3. Checking individual offset

Input: Validated pair of inliers, Image width
Output: Individual offset alignment
1: Sum of individual offsets:
$$\text{for } i = 0 \longrightarrow n = lenght(inliers\_pairs)$$
$$\text{num}(i) = \frac{second\_panorama\_inlier(i) - first\_panorama\_inlier(i)}{Image\_width} \cdot 360;$$
$$final\_num = final\_num + num(i);$$
2: $Individual\_offset\_alignment \longleftarrow offset\_angle = \frac{final\_num}{n};$

---

under established tolerance in relation with angle and module conditions. In order to validate the theoretical purged inliers after M-SAC filter, all keypoints vectors must be validated. Otherwise, even just one of evaluated inliers is defective, the combination among detection techniques (Harris, SURF and SIFT) and different homogeneous transformation applying the M-SAC filter (similarity, affine or projective) will do the inlier is automatically discarded. It can be summarized on a 3 (detection techniques) x 3 (homogeneous transformations) quality matrix. If the combination was discarded by the parallel filter, the position will be zero in the matrix. In any case, this matrix element will store the number of resulting pairs of inliers after applying M-SAC filter.

When the matrix is totally filled, it is ordered from highest to lowest number of inliers, considered in this work as the main quality parameter. The greater the number of matched pairs, the greater the accuracy of the estimated transformation. If exist four or more elements non-zero, there will be chosen the three highest values. A great checked matching example is observed in Fig. 7. Notice that the number of keypoints in the homogeneous transformation stage tends to be larger due to the extraction is carried out now with both images horizontally aligned as far as possible. Figure 8 shows an example of progressive alignment. First and second row show original first and second synthesized panorama views. Third row shows second synthesized panorama view after horizontal alignment and fourth row shows second synthesized panorama views after applying homogeneous transformation when required. Be aware that middle column (second pair of images) has such amount of change in point-of-view that neither the global alignment offset nor individual alignment offset by using keypoints can correct the initial yaw angle, so this pair of images would be discarded.

2.6 Differential Groundtruth

Labeling represents the last stage of our application. By using Liblabel Matlab tool created by Autonomous Vision Group of the University of Tübingen [18], we are able to label in an easy way structural changes that span from road repairements to facades, traffic signals or construction/demolition of new buildings. This lasts interface lets the user adding/removing a label or even generating the groundtruth at the same moment in order to check if a right semantic segmentation has been carried out.

Fig. 7. Recommended alignment interface. It can be observed that results are completely coherent and successful.

Algorithm 4. Validation by using parallel filter

Input: Estimated keypoints pairs
Output: Alignment validation

1: Resulting keypoints pairs after M-SAC filter are taken.
2: Reference point calculation:

  k(1)=(second.keypoints.Location.x + 640)-(first.keypoints.Location.x);
  l(1)=(second.keypoints.Location.y + 640)-(first.keypoints.Location.y);

3: The first pair represents the reference absolute distance and reference angle:
$$module(1) = \sqrt{k(1)^2 + l(1)^2};$$
$$angle(1) = \frac{360}{2\cdot\pi} \cdot \arcsin(\frac{l}{module(1)});$$

4: Alignment validation depending on manual set tolerances:
$$\text{for } i=2 \longrightarrow index(n)$$
$$module(i) = \sqrt{k(i)^2 + l(i)^2};$$
  If (module(i) inside module tolerances then $\longrightarrow Calculate\_angle(i)$;
  If (abs_angle inside angle tolerances) then $\longrightarrow cont++$;

Notice that in this work we will label in a mono-class segmentation way: There is structural change (taking into account its several classes) or not, not distinguishing in several classes which would correspond to a multi-class segmentation. Figure 9 shows main structural changes labeled throughout this work, being the three most representative classes: Repairements/maintenance (36%), large structural changes (21%) and combination among different classes (21%). To a lesser extent, facades (10%), non-presence of changes (9%) and singular changes in traffic signals (3%) are found. It can be observed that the percentage of repairements is bigger than others because there were more structural changes related to demolition/construction of buildings rather than traffic signals.

An important contribution in this work is the incorporation of a link between the tool and output database that contains the polygon coordinates of those pair of images from where the groundtruth has already been generated. Thus, if we

Fig. 8. Progressive alignment results performed by our application.

try to label a pair of images that were previously labeled in order to improve the quality of the groundtruth or even labeling to a complex multi-class segmentation, last information remains unless it is removed, speeding up the process.

3   Validation Results

So as to validate our proposed tool and the dataset obtained with it, we trained and validated a DNN proposed by one of the authors and we compare its performance with other works of the state-of-the-art. First, we divided the obtained instances dataset into a training set of 844 instances and a validation set of 233 instances. The split among sequences was chosen at random, with whole image descriptor matching used to confirm that test and training sets did not contain similar looking sequences.

In order to elaborate this training/validation instances dataset, 559 geolocation links where collected all over the world, then processed 354 of them by using the tool (as illustrated in Table 1) with the intention of generating a total of 354 · 2 · 3 = 2,124 synthetic images, because each link represents the same place in different time. From each place three synthetic views are obtained, as shown throughout this work. Notice that is much more convenient talking about 354 · 3 = 1,062 images pairs rather than 2,124 synthetic images because an image pair is the main element of each training/validation instance. Those image pairs that presented strong misalignment in terms of rotation and translation, even after global and individual offset, were discarded in order not to introduce unnecessary noise to DNN. So, from a total of 1,062 theoretical training/validation instances we got 909 effective instances (as average, 4 min distributed in analyzing and labeling each pair) in order to train and test the DNN, obtaining an 85.59 %

of efficiency. In the instances obtaining process our dataset presents an order of magnitude larger that previous databases of the state-of-the-art, as shown in Table 1.

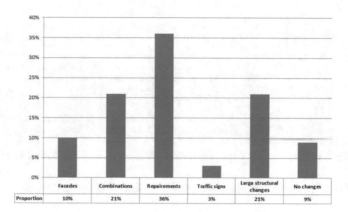

Fig. 9. Main structural changes detected. Most of them are repairements/maintenance tasks and secondly medium size structural changes.

Table 1. Comparison of existing Street-View Change Detection Datasets

Dataset	#Sequences	#Pairs	#Type
Taneja et al. [12]	4	50	Perspective
Sakurada et al. [11]	23	92	Perspective
Sakurada and Okatani [19]	-	200	Panoramic
VL-CMU-CD [7]	152	1,362	Perspective
Current Dataset (Ours)	559 (354 used)	909	Perspective

In addition, with the purpose of helping prevent overfitting during training we used data augmentation by adding pairs of images containing both no changes of interest and artificial changes (by adding synthetic changes to existing images) [20]. This place recognition method has shown [21] impressive performance to recognize images from the same place under different weather and lighting conditions, taking into account that there are no large structural changes between both images, providing a pretty useful source of data augmentation for structural change detection systems.

These additional images were added in an approximate ratio of 25% augmented changes and 75% real changes, presenting the training dataset a total amount of 909 + 300 = 1,209 instances.

We compared our DNN called CDNet against multiple state-of-the-art techniques: Hand-crafted descriptors such as DAISY [10], DASC [8] (recently introduced as dense descriptor for multi-spectral and multi-modal correspondences)

and dense SIFT [9]. Furthermore, our CDNet is compared to a pre-trained CNN for image recognition combined with superpixel regularization and sky segmentation [19].

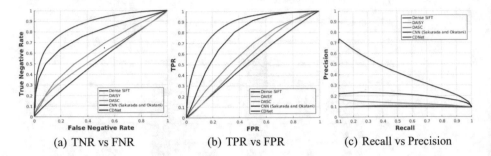

(a) TNR vs FNR   (b) TPR vs FPR   (c) Recall vs Precision

Fig. 10. Validation test results after applying the architecture of the DNN to our dataset

Quantitative Comparison: Figure 10 shows a comparison of the CDNet based on our instances obtaining process versus previous baseline methods. We show *True Negative Rate (TNR) or specificity* versus *False Negative Rate (FNR)*, *True Positive Rate (TPR) or Recall* versus *False Positive Rate (FPR)* and *Precision (Pr)* versus *Recall (Re)* graphs. These are defined as follows: TPR or Recall: TP/(TP+FN), FPR: TP/(TP+FN), Pr: TP/(TP+FN) and f1_Score: $2 \cdot Pr \cdot Re/(Pr + Re)$.

Table 2. Quantitative comparison versus baseline methods at FPR = 0.10

Method	Pr	Re	F1-Score
Dense SIFT [9]	0.1133	0.1078	0.1105
DAISY [10]	0.1112	0.1063	0.1087
DASC [8]	0.1944	0.1799	0.1868
CNN (Sakurada and Otakani [19])	0.3671	0.3000	0.3302
CDNet (Ours)	**0.6078**	**0.5889**	**0.5982**

Table 2 compares our method over different change detection metrics for a FPR of 0.10. This same value was used in the qualitative comparison, as illustrated in Fig. 11. Notice that our improved DNN architecture outperforms other methods on all metrics by a significant margin, being these best results highlighted in bold.

Qualitative Comparison: Figure 11 shows qualitative results after applying our improved DNN to validate instances generated by the tool, with a FPR of 0.10. It is illustrated the predicted change maps of some methods for randomly

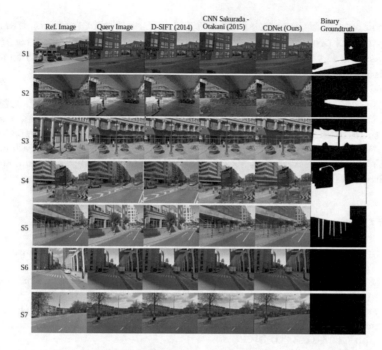

Fig. 11. Structural changes detection by using validation instances and different approaches. Changes detected are labelled in red.

selected validation image pairs. The performance of the DNN over these valida-tion instances generated by the tool is quite better than other methods, both in detecting when there is structural change (from S1 to S5 row in Fig. 11) and no detecting when there is no structural change (S6 and S7 in Fig. 11), what shows both great performance of the network and great performance of the tool in the task of training and validating the DNN. Notice that changes due to vegetation, pedestrians or vehicles are not labeled.

4   Conclusions and Future Works

We have created a tool able to process links from Internet with GSV and generate aligned pair of images from the same geolocation but different time situations as well as a binary groundtruth able to detect structural differences between the pair of images. In addition, we create a new dataset in order to train and validate a DNN able to detect street-view structural changes. Our method combines panoramic images and depth maps generation, view synthesis, image alignment and manual labeling. Taking into account that the efficiency of our tool is 85.6 %, in relation to the searching and processing time of the dataset (2 months) results can be considered successful.

Future improvements are the generation of our own depth map from the difference of pose between two panoramas of the same temporal situation and same geolocation, an improved camera pose correction in terms of translation in the synthesis process, the development of a better alignment process in order to increase the DNN labeling efficiency and the geolocation database to at least ten thousand of links.

Acknowledgment. This work has been partially funded by the Spanish MINECO/ FEDER through the SmartElderlyCar project (TRA2015-70501-C2-1-R), the DGT through the SERMON project (SPIP2017-02305), and from the RoboCity2030-III-CM project (Robótica aplicada a la mejora de la calidad de vida de los ciudadanos, fase III; S2013/MIT-2748), funded by Programas de actividades I+D (CAM) and cofunded by EU Structural Funds.

References

1. Martin-Brualla, R., Gallup, D., Seitz, S.M.: Time-lapse mining from internet photos. ACM Trans. Graph. (TOG) **34**(4), 62 (2015)
2. Matzen, K., Snavely, N.: Scene chronology. In: European Conference on Computer Vision, pp. 615–630. Springer (2014)
3. Krizhevsky, A., Sutskever, I., Hinton, G.E.: Imagenet classification with deep convolutional neural networks. In: Advances in Neural Information Processing Systems, pp. 1097–1105 (2012)
4. Noh, H., Hong, S., Han, B.: Learning deconvolution network for semantic segmentation. In: Proceedings of the IEEE International Conference on Computer Vision, pp. 1520–1528 (2015)
5. Sünderhauf, N., Shirazi, S., Jacobson, A., Dayoub, F., Pepperell, E., Upcroft, B., Milford, M.: Place recognition with convnet landmarks: viewpoint-robust, condition-robust, training-free. In: Proceedings of Robotics: Science and Systems XII (2015)
6. Ros, G., Stent, S., Alcantarilla, P.F., Watanabe, T.: Training constrained deconvolutional networks for road scene semantic segmentation, arXiv preprint arXiv:1604.01545 (2016)
7. Alcantarilla P.F., Stent, S., Ros, G., Arroyo, R., Gherardi, R.: Street-view change detection with deconvolutional networks. In: Robotics: Science and Systems (2016)
8. Kim, S., Min, D., Ham, B., Ryu, S., Do, M.N., Sohn, K.: DASC: dense adaptive self-correlation descriptor for multi-modal and multi-spectral correspondence. In: Proceedings of the IEEE Conference on Computer Vision and Pattern Recognition, pp. 2103–2112 (2015)
9. Lowe, D.G.: Distinctive image features from scale-invariant keypoints. Int. J. Comput. Vis. **60**(2), 91–110 (2004)
10. Tola, E., Lepetit, V., Fua, P.: DAISY: an efficient dense descriptor applied to wide-baseline stereo. IEEE Trans. Pattern Anal. Mach. Intell. **32**(5), 815–830 (2010)
11. Sakurada, K., Okatani, T., Deguchi, K.: Detecting changes in 3D structure of a scene from multi-view images captured by a vehicle-mounted camera. In: 2013 IEEE Conference on Computer Vision and Pattern Recognition (CVPR), pp. 137–144. IEEE (2013)
12. Taneja, A., Ballan, L., Pollefeys, M.: Image based detection of geometric changes in urban environments. In: 2011 IEEE International Conference on Computer Vision (ICCV), pp. 2336–2343. IEEE (2011)

13. Finman, R., Whelan, T., Kaess, M., Leonard, J.J.: Toward lifelong object segmentation from change detection in dense RGB-D maps. In: 2013 European Conference on Mobile Robots (ECMR), pp. 178–185. IEEE (2013)
14. Cavallo, M.: 3D city reconstruction from google street view. Comput. Graph. J. (2015)
15. Torii, A., Arandjelovic, R., Sivic, J., Okutomi, M., Pajdla, T.: 24/7 place recognition by view synthesis. In: Proceedings of the IEEE Conference on Computer Vision and Pattern Recognition, pp. 1808–1817 (2015)
16. Harris, C., Stephens, M.: A combined corner and edge detector. In: Alvey Vision Conference, vol. 15, pp. 10–5244. Citeseer (1988)
17. Wang, H., Mirota, D., Hager, G.D.: A generalized kernel consensus-based robust estimator. IEEE Trans. Pattern Anal. Mach. Intell. **32**(1), 178–184 (2010)
18. Geiger, A., Lauer, M., Wojek, C., Stiller, C., Urtasun, R.: 3D traffic scene understanding from movable platforms. In: Pattern Analysis and Machine Intelligence (PAMI) (2014)
19. Sakurada, K., Okatani, T.: Change detection from a street image pair using CNN features and superpixel segmentation. In: BMVC, pp. 1–61 (2015)
20. Stent, S., Gherardi, R., Stenger, B., Cipolla, R.: Detecting change for multi-view, long-term surface inspection. In: BMVC, pp. 1–127. Citeseer (2015)
21. Arroyo, R., Alcantarilla, P.F., Bergasa, L.M., Romera, E.: Fusion and binarization of CNN features for robust topological localization across seasons. In: 2016 IEEE/RSJ International Conference on Intelligent Robots and Systems (IROS), pp. 4656–4663. IEEE (2016)

Improving the 3D Perception of the Pepper Robot Using Depth Prediction from Monocular Frames

Zuria Bauer, Felix Escalona$^{(\boxtimes)}$, Edmanuel Cruz, Miguel Cazorla,
and Francisco Gomez-Donoso

Institute for Computer Research, University of Alicante,
P.O. Box 99, 03080 Alicante, Spain
felix.escalona@ua.es

Abstract. The robot Pepper provides a bad depth estimation. In this paper, we present a method for improving that 3D estimation. The method is based on using the RGB image to predict monocular depth. As it will be shown, the combination of both, monocular and 3D depth, provides a better 3D data.

Keywords: Depth perception · Pepper robot · Depth estimation
Sensor fusion

1 Introduction

In recent years, it is undeniable that there has been a growing interest in social robotics. This fact is reflected in the robotic industries that have developed robotic products focused on social interaction. As an example of this trend, we can refer to the Pepper Robot, which is a humanoid robot manufactured by SoftBank Robotics (formerly Aldebaran Robotics). This robot was intended to provide a friendly companion with emotional analysis capabilities at home and office environment. As expected, many research groups worldwide have developed a large amount of projects based on this robotic platform due to its features. This robot has a large number of sensors and hardware capabilities oriented to solve indoor tasks such as navigation, scene understanding and user interaction. The robot incorporates an Asus Xtion 3D sensor in its 1.8a hardware version. This camera is widely used by researchers as a standalone device, however the version included in the robot is not working properly. As a result, the tridimensional representation of the environment provided by the robot is quite inaccurate and noisy, rendering it nearly useless.

As a social robot, one main task that is intended to perform is the navigation. A variety of state-of-the-art navigation and SLAM algorithms, such as [6,10] or [11], rely on point clouds. As the depth camera is not working properly, it would be arduous to implement those methods on the robots.

© Springer Nature Switzerland AG 2019
R. Fuentetaja Pizán et al. (Eds.): WAF 2018, AISC 855, pp. 132–146, 2019.
https://doi.org/10.1007/978-3-319-99885-5_10

In addition to this problem, which is inherent only to the 1.8a version of the Pepper Robot, the structured light based sensors also have another issues when it comes to the tridimensional representation of the environment.

In this paper, the main proposal aims for a method to improve the depth perception of the Pepper Robot v1.8a in order to generate more accurate 3D representations. Our proposal successfully fuses the depth maps provided by the robot with a predicted depth map generated by a depth estimation from monocular frames, which is a deep learning based approach. In addition, our proposal also tackles another problems that affect all the structured light and time of flight based sensors. Exhaustive experimentation validate that our system improves the Pepper Robot v1.8a depth perception.

2   Related Works

This work tries to solve some problems of the current tridimensional sensors by fusing the point clouds provided by an actual depth camera and the estimated depth map of a deep leaning based approach. So, this section is split in these three corresponding subsections.

2.1   Depth Cameras Issues

Researchers in many areas including human-computer interaction (HCI), computer vision, robotics and augmented reality (AR), use depth sensor for their researches. However, these sensors have different issues when it comes to the tridimensional representation of the environment. The factors related to noise in depth maps have been investigated from various perspectives.

In [20] They characterize the noise in Kinect depth images based on multiple factors and introduce a uniform nomenclature for the types of noise. They observe that four parameters, namely, *object distance, imaging geometry, surface/medium property*, and *sensor technology* control spatial noise. They based on these four control parameters and the source of noise to characterize the noise behavior into *out-of-range, axial, shadow, lateral, specular surface, non-specular surface, band, structural*, and *residual noise* classes. For each class, they summarize the noise behavior either from others' reports or from their experiments. Object distance, motion, surface properties, and frame rate control the behavior of temporal noise. Hereafter we describe some Depth camera issues using this classification.

The *shadow noise* [27] is produced by objects that obstruct the path of infra-red (IR) emitter or IR camera or both. The characteristics of this noise are: elongated component following object boundary, zero depth, nearly uniform in width, decreases with increasing distance from the sensor, consistent with IR lighting direction, decreases with a near-by background.

The *lateral noise* [22] is produced by shadow-like non-uniformity at edges. The characteristics of this noise are pronounced for straight, especially, vertical edges, Error along edges, varies linearly with depth, Zero Depth, decrease with nearby background and occurs at both shadow and non-shadow edges.

The *specular surface noise* [20] is produced by highly IR reflecting surface fails to diffuse the speckle pattern. The characteristics of this noise are consistent across frames, zero depth could be irregular big patches of zero Depth not adjacent to the image border. Distant (reflected) objects may be recorded at WD.

2.2   Sensor Fusion Techniques

In nature, it is common to see a species combining signals to recognize its environment. Humans have the ability to recognize their environment by combining information from multiple sensors such as ears, tongue, skin, and eyes. Today the application of fusion concepts in technical areas has constituted a discipline, that spans many fields of science [5]. Sensor fusion offers a great opportunity to overcome physical limitations of sensing systems. In the literature, there are many definitions for data fusion.

Castanedo [2] defined data fusion as a combination of multiple sources to obtain improved information; in this context, improved information means less expensive, higher quality, or more relevant information.

Elmenreich [5] describe Sensor Fusion as the combining of sensory data or data derived from sensory data such that the resulting information is in some sense better than would be possible when these sources were used individually. According to [2] the data fusion techniques can be classified into three non-exclusive categories: (i) data association, (ii) state estimation, and (iii) decision fusion.

Data Association Techniques [2]. The aim of data association is to establish the set of observations or measurements that are generated by the same target over time. The most common techniques that are employed to solve the data association problem in this category are *Nearest Neighbors* and *K-Means*, *Probabilistic Data Association, Joint Probabilistic Data Association, Multiple Hypothesis Test, Distributed Joint Probabilistic Data Association, Distributed Multiple Hypothesis Test* and *Graphical Models*.

State Estimation Methods [2]. State estimation techniques aim to determine the state of the target under movement (typically the position) given the observation or measurements. The most common estimation methods are *maximum likelihood* and *maximum posterior*, the *Kalman filter*, *particle filter*, the *distributed Kalman filter*, *distributed particle filter* and *covariance consistency methods*.

Decision Fusion Methods [2]. A decision is typically taken based on the knowledge of the perceived situation, which is provided by many sources in the data fusion domain. The most common Fusion Methods are the *bayesian methods*, the *Dempster-Shafer Inference, Abductive Reasoning* and *Semantic methods*.

The selection of the most appropriate technique depends on the type of the problem and the established assumptions of each technique.

2.3   Depth Estimation from Monocular Frames

Depth estimation from image data has been studied for a long time, originally relying on stereo vision [25]. In this section we will review only the monocular centered methods organized in chronological order.

The first work was published in 2005 by Saxena [24]. The paper presented an approach to depth estimation from a single monocular image with an supervised learning approach. The model used discriminative-trained Markov Random Field (MRF) that incorporated local and global features from the images.

Years later, in 2010, Liu et al. published a paper [17] using semantic segmentation of the scene and guiding the 3D reconstruction with it. The model also used MRF in order to enforce neighboring constraints.

Another interesting paper was [9], which presented a depth transfer approach in three stages. First, they searched for the candidate images. Secondly, they applied a warping procedure to the candidate and depth images. Finally, they interpolated and smoothed the warped candidate depth values; these results where the inferred depth.

The same year was released [12], whose approach relied on the observation that from all the pairs of image+depth available on-line, there likely existed many pairs whose 3D content matched a 2D input (query).

In 2014, Eigen published one of the most remarkable papers in this area [4]. They used a coarse-scale network to predict the depth of the scene at a global level. This was then refined within local regions by a fine-scale network. Both stacks were applied to the original input, but in addition, the coarse network's output is passed to the fine network as additional first-layer image features.

The same year was published [19]. Their goal was to estimate the depth of the pixels observed in a single image depicting a general scene. They used the terms of superpixels to approach this topic, making the common assumption that each superpixel is planar.

In 2015, there were three different papers published in this area:

The first was [18]. They used a deep structured learning scheme which learns the unary and pairwise potentials of continuous CRF in a unified deep CNN framework.

The second was [14]. The paper proposed a two step model. First step was the use of a DCNN model to learn the mapping from multi-scale image patches to depth. Second, the estimated super-pixel depth or surface normal is refined to the pixel level by exploiting various potentials on the depth or surface normal map.

The third paper was [26]. They used a trained Convolutional Neural Network (CNN) to jointly predict a global layout composed of pixel-wise depth values and semantic labels.

The following year, Laina et al. published [13], that we have used as baseline for this article. They applied a fully convolutional architecture to predict depth, endowed with novel up-sampling blocks, that allowed for dense output maps of higher resolution and at the same time required fewer parameters.

In 2016, the article [23] was published. It combined CNNs with Regression Forest for predicting depths in the continuous domain via regression. They achieved robustness by processing a data sample with an ensemble of binary regression trees, which they called Convolutional Regression Trees (CRTs).

Also that year appeared the paper [1]. It took as input an arbitrarily sized image and outputs a dense score map. Fully connected CRFs are then applied to obtain the final depth estimation.

Other paper was [16]. It captured scene details by considering information contained in depth gradients. It postulated that local structure encoded with first-order derivative terms.

Another important paper was the [15]. It aims at predicting pixel-wise depth from a single color image. They propose a simple and effective dilated deep residual CNN architecture, which could converge with much fewer training examples and model parameters.

3   Pepper Robot

As aforementioned, this paper aims to provide a solution to the issue that affects the 3D camera of the Pepper Robot, but first we are briefly describing the Pepper Robot features.

Pepper is a humanoid robot developed by Aldebaran Robotics designed as a social robot. It is intended to provide a human-like interaction with its speech and gesture capabilities. It was presented the 5th June 2014 in Urayasu, Tokyo. The dimensions of the robot are as follows: it features a height of 1.21 m, 0.48 m of width and 0.425 m of depth. It features a variety of sensors, from microphones to laser, radar and color cameras, as it can be seen in Fig. 1.

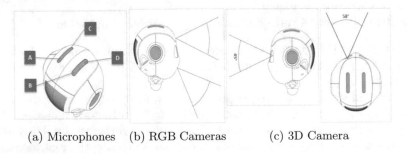

(a) Microphones   (b) RGB Cameras   (c) 3D Camera

Fig. 1. Some sensors featured by the Pepper Robot. (a) depicts the position of the microphones array, (b) shows the field of view and position of both color cameras, and in (c) the 3D camera is displayed.

There are currently three different hardware versions of the robot. The version 1.8, which is the most recent one, comes with a totally new tridimensional sensor. This version of the robot has two 4 MPx color cameras, located in both "eyes", that provide depth perception using a stereo algorithm. The resultant depthmaps are 1280×720 resolution and are generated at 15 frames per second. Nonetheless, versions 1.8a and 1.6 come with the aforementioned faulty Asus Xtion depth sensor. This sensor is able to provide 320×240 resolution depthmaps at 20 frames per second. Further information about the hardware versions and the depth cameras can be found on the official documentation site[1].

3.1   The Problem with the Tridimensional Sensor of the Pepper Robot

As aforementioned, the Pepper Robot includes an Asus Xtion sensor as the depth camera. This camera is theoretically able to provide decent depthmaps of the scenes. Nonetheless, the Xtion that is equipped with the robot seems to provide erroneous depthmaps that lead to also erroneous point clouds.

As it can be seen in Fig. 2, this problem consists in a distorted representation of the tridimensional space. The resultant point clouds show a wave-like pattern across all the scene. It becomes more evident when it depicts plane artifacts, such as walls or floors, but it exists in all the scene. In addition, we noticed that the distortion worsen as the depth increases, namely, the objects that are near the sensor are less affected than those that are at greater distances.

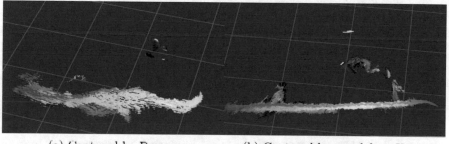

(a) Captured by Pepper   (b) Captured by standalone Xtion

Fig. 2. Pepper Robot's Xtion camera and a external standalone Xtion point clouds. Note that the images depict the same scene and are taken from the same point-of-view. The white artifact is actually a wall, which is a planar surface.

In order to quantitative evaluate the distortion of the Pepper depth camera, we carried out a simple experiment. We took a planar object and set it at different distances from the 3D sensor of the Pepper Robot. The distances ranged from 1 to 3 m, increasing the distance 0.5 m. For each case, we captured the corresponding depth map and projected it to the 3D space. Then, we manually selected all the points laying in the planar object. This step is straightforward as the depth maps and color frames are registered. Then, we fitted a plane using RANSAC [7] setting the inlier threshold to infinite. This is done to ensure that all the points will be inliers. Due to the random nature of RANSAC, 60000

---

[1] http://doc.aldebaran.com/2-7/family/pepper_technical/video_3D_pep.html.

different planes were tested but only the one with the lowest RMSE is returned. Results of these experiments can be found in Fig. 3. The mean euclidean distance from each point to the estimated plane is reported for each experiment in the figure.

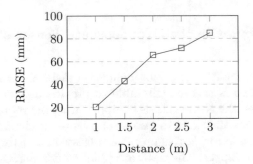

Fig. 3. Mean error per distance to plane achieved by the Pepper Robot.

In the light of the results, it can be stated that the mean error grows as the distance to the object is increased. Namely, the quality of the resultant point clouds gradually worsen. Speaking in absolute terms, the 3D representation of the planar object is awful even when the object is as near as 1 m. In this case, the mean error is 20.36 mm, but the furthest point is 79.15 mm from the plane. The furthest point in the 3 m experiment is 279.12 mm. Figure 4 depicts the points clouds and corresponding estimated planes for 1 and 3 m for qualitative evaluation purposes. Note that the version of NaoQi and other methodological details related to Pepper we used in this experiment can be found in Sect. 5.

Finally, it is worth noting that ours is not an isolated faulty unit. We have contacted personally with fellow researchers from other universities and laboratories and they have also reported the same issue with the same robot version.

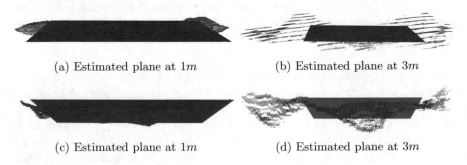

(a) Estimated plane at 1m   (b) Estimated plane at 3m

(c) Estimated plane at 1m   (d) Estimated plane at 3m

Fig. 4. Estimated plane for the point cloud representation of a planar object. Note that the representation is far from accurate in both cases. Furthermore, the issue worsen as the distance is increased.

3.2   Other Issues Related to 3D Cameras

There are other issues that commonly affects all the structure light based and time of flight based sensors. These technologies are used by most commercial depth cameras such as the Asus Xtion and Microsoft Kinect. The first issue we are discussing is the impossibility to compute the depth in the shadow that cast the foreground objects in the background objects. This case produces some areas with no depth information around the boundaries of the objects in the scene. This case can be seen in Fig. 5a. As shown, some pixels of the depth map yield no valid depth values.

Another issue is the incompatibility with specular surfaces. Objects with specular surfaces are not sensed by those sensors, and produces areas with no depth information or faulty depth values in the resultant depth maps. This case can be seen in Fig. 5b. As shown, the depicted surface is specular so the depth map does not represent it properly.

(a) Shadow of foreground objects   (b) Specular surfaces

Fig. 5. Other issues related to 3D cameras

4   Fusing the Output of a Depth Estimation from Monocular Frames System and the Pepper Depth Maps

We propose the fusion of the erroneous point cloud provided by the Pepper Robot with a depth estimation from monocular frames method. First, the point cloud provided by the Pepper Robot is obtained as is. No further preprocessing is performed on this data.

Second, the method proposed in [13] is used to estimate the depth map from a color image. This system describes a fully convolutional neural network that is able to predict depth maps taking a single color image as input. The methodology provided in the original paper was adopted to perform this experiment.

The architecture Fig. 6 features a fully convolutional neural network build upon a ResNet50 [8]. This architecture includes different convolution and pooling blocks with eventual residual connections followed by a last fully connected layer that was first presented as a classifier for the ImageNet challenge. In this incarnation, the last fully connected layer is replaced by a number of Up-Projection layers.

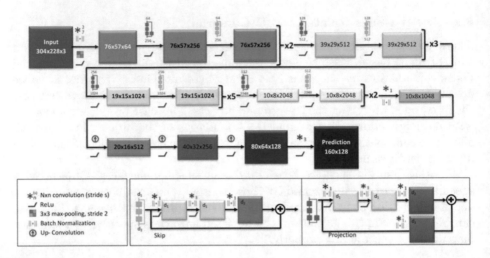

Fig. 6. The benchmarked approach is based on a ResNet50. In this case, the last fully connected layer was replaced by several upsampling layers.

These Up-Projection blocks are presented as the main novelty of this architecture. They are based on the un-pooling method proposed in [3], but they extended the idea introducing residual and projection connections. By chaining up to four of these blocks, this architecture achieves efficient high-level features forwarding while increasing the resolution of the tensors.

Note that the architecture was trained with the NYU Dataset [21] and takes 304×228 resolution color images as input and predicts 160×128 depth maps, which are resized to fit the original image size to allow straightforward align with the color image and the 2D object detections.

Once we obtained the depth map from the Pepper Robot and the estimated depth map by the aforementioned method, both data are summarized. The summarization is performed as follows:

1. **Look for NaN points in Pepper point cloud.** Find invalid points in Pepper point cloud and save their indexes for further processing. Those points correspond to loss of depth information produced by the aforementioned shadow effect and specular surfaces.
2. **Compare points distances between Pepper and Predicted point clouds.** Each point of the Pepper point cloud is compared with the Predicted point cloud and the distances are computed. This process is straight forward as both points clouds are registered. If the distance is over a threshold Te, the point of the Pepper point cloud is inserted in the Fused point cloud. We do that in order to avoid worsen the Fused output with erroneous predictions.
3. **Calculate the planarity of the remaining points in monocular estimation.** For each remaining point, we calculate a planarity measure. First, we search its neighbor points within a radial threshold r. Then, we approximate

a plane using RANSAC, which is robust against outliers. Finally, we calculate the ratio of neighbor points whose distance to the plane is lower than a certain threshold d. This is the planarity measure. If this measure is higher than a determined value Tp, then this information is inserted in the output Fused point cloud. Otherwise, the output information is taken from Pepper point cloud.

4. **Fill the gaps in the output point cloud.** Using the indexes we took apart in the first step, we pick the corresponding points in the Predicted point cloud and check if they are consistent with their neighbors within a certain range in the output ro. If the median of the distance with their neighbors is lower than a certain threshold m, then we add this information to the resultant Fused point cloud.

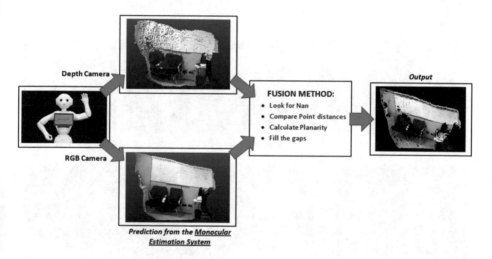

Fig. 7. Scheme of the proposed method for point cloud fusion

The result is a corrected point cloud where the planes are enhanced and the density preserved due to the utilization of the Predicted point cloud, while we avoid the artifacts and isolated erroneous depths produced by the Predicted point cloud as much as possible.

5   Experimentation

Before the presentation of the experiments we carried out and the corresponding discussion, it is worth stating the hardware setup for reproducibility purposes.

First, our Pepper Robot test unit is version 1.8a. In order to operate the robot, it is mandatory to use the NaoQi API provided by SoftBank. All the experiments were carried out with the last version of the NaoQi, which is the version 2.8 as of today. The tridimensional sensor in this unit is an Asus Xtion. The focal length is 525, the center of the image in the X axis is 319.5 whist the center of the image in the Y axis is 239.5. The parameters were provided by the manufacturer and were not estimated. As this sensor provides depth maps, we used the described intrinsic parameters to project the depth data to the 3D space.

Regarding the hardware specifications of the test setup, we used a Intel Core i5-3570 with 8 GiB of Kingston HyperX 1600 MHz and CL10 DDR3 RAM on an Asus P8H77-M PRO motherboard (Intel H77 chipset). The system also included a NVIDIA Quadro P6000 GPU. The framework of choice was Keras 1.2.0 with Tensor Flow 1.8 as the backbone, running on Ubuntu 16.04. CUDA 9.0 and cuDNN v5.1 were also used to accelerate the computations.

In order to demonstrate the reliability of the proposed method, we have made a set of captures of a whiteboard of our lab at different distances, as seen in Fig. 8.

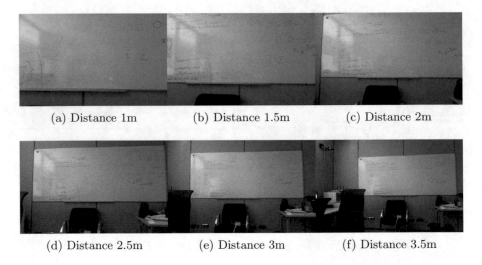

(a) Distance 1m   (b) Distance 1.5m   (c) Distance 2m

(d) Distance 2.5m   (e) Distance 3m   (f) Distance 3.5m

Fig. 8. RGB images of the scenes for plane comparisons between Pepper and Fusion point clouds.

Table 1. Point to plane distances (Root Mean Squared Error) of Pepper and Fusion point clouds.

Distance (m)	RMSE Pepper (mm)	RMSE Fusion (mm)
1.0	10.7442	17.6085
1.5	19.8922	10.1363
2.0	37.829	24.3125
2.5	54.4989	37.2627
3.0	73.4166	53.571
3.5	126.011	44.1466

Then, we have made a comparison between the point clouds generated by Pepper and by our fusion method. We have taken the whiteboard planes from every scene and calculated the RMSE of every point inside them, as shown in Table 1. Processing times aren't shown because the processing of our fusion method is being made in a external computer and it has not been optimized yet

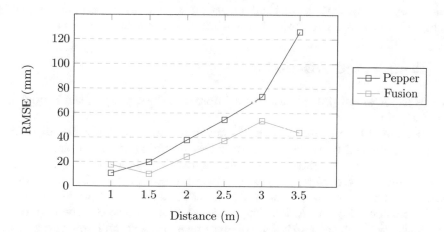

Fig. 9. Point to plane distances (Root Mean Squared Error) of Pepper and Fusion point clouds.

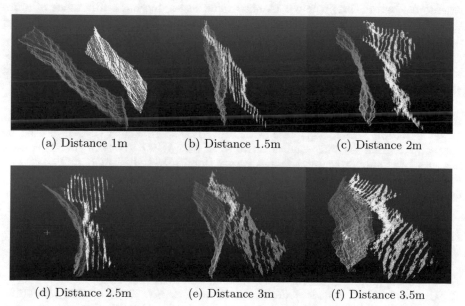

(a) Distance 1m   (b) Distance 1.5m   (c) Distance 2m

(d) Distance 2.5m   (e) Distance 3m   (f) Distance 3.5m

Fig. 10. Plane comparisons between Pepper (white) and Fusion (gray) point clouds. Pepper planes worsen as the distance to the camera increases.

with parallelization techniques. We have focused on the reliability of our method rather than its speed.

The results, presented as a plot in Fig. 9, show that the RMSE of the Pepper planes increases as a function of the distance. On the other hand, the RMSE of our fusion method, based on monocular estimation, is more dependent on the texture and environment information than on camera distance. Comparing both graphs, we can see that our method outperforms the original point cloud quality. In Fig. 10 we can see the projection of every plane of this experiment.

6 Results and Discussion

In this paper we propose a method for improving the quality of the Pepper 1.8a depth map.

As exposed before, the camera of Pepper suffers a kind of radial distortion, arguably produced by its lens, that produces several artifacts on planes.

We profit the advantages in monocular depth estimation in order to have a depth map that performs better in plane representation. This estimation is made with the *Iro Laina* architecture, one of the state of the art methods.

After calculating this estimation, we make a fusion between the Pepper and monocular point cloud, looking for planar areas in the second one, and generate a corrected point cloud.

As shown in the Experimentation section, the fused output point cloud represents far better the principal planes in the scene, having a difference of more than 80 mm on the RMSE for 3.5 m. This represents a great improvement and demonstrates the reliability of the proposed method.

For future work, we are focusing on improving the distance estimation provided by the monocular method and investigating smarter ways of depth fusion considering the pro and cons of every method.

Acknowledgments. This work has been supported by the Spanish Government TIN2016-76515R Grant, supported with Feder funds. Edmanuel Cruz is funded by Panamenian grant for PhD studies IFARHU & SENACYT 270-2016-207. This work has also been supported by a Spanish grant for PhD studies ACIF/2017/243 and FPU16/00887. Thanks to Nvidia also for the generous donation of a Titan Xp and a Quadro P6000.

References

1. Cao, Y., Wu, Z., Shen, C.: Estimating depth from monocular images as classification using deep fully convolutional residual networks, May 2016
2. Castanedo, F.: A review of data fusion techniques. Sci. World J. **2013**, 19 (2013)
3. Dosovitskiy, A., Springenberg, J.T., Tatarchenko, M., Brox, T.: Learning to generate chairs, tables and cars with convolutional networks. arXiv e-prints, November 2014

4. Eigen, D., Puhrsch, C., Fergus, R.: Depth map prediction from a single image using a multi-scale deep network. In: Proceedings of the 27th International Conference on Neural Information Processing Systems - Volume 2, NIPS 2014, pp. 2366–2374. MIT Press, Cambridge (2014). http://dl.acm.org/citation.cfm?id=2969033.2969091

5. Elmenreich, W.: An introduction to sensor fusion. Research Report 47/2001, Technische Universität Wien, Institut für Technische Informatik, Treitlstr. 1-3/182-1, 1040 Vienna, Austria (2001)

6. Engelhard, N., Endres, F., Hess, J., Sturm, J., Burgard, W.: Real-time 3D visual SLAM with a hand-held RGB-D camera. In: Proceedings of the RGB-D Workshop on 3D Perception in Robotics at the European Robotics Forum, Vasteras, Sweden, April 2011

7. Fischler, M.A., Bolles, R.C.: Random sample consensus: a paradigm for model fitting with applications to image analysis and automated cartography. Commun. ACM **24**(6), 381–395 (1981). https://doi.org/10.1145/358669.358692

8. He, K., Zhang, X., Ren, S., Sun, J.: Deep residual learning for image recognition. arXiv e-prints, December 2015

9. Karsch, K., Liu, C., Kang, S.B.: Depth extraction from video using non-parametric sampling. In: Proceedings of the 12th European Conference on Computer Vision - Volume Part V. ECCV 2012, pp. 775–788. Springer, Heidelberg (2012). https://doi.org/10.1007/978-3-642-33715-4_56

10. Kim, P., Chen, J., Cho, Y.K.: SLAM-driven robotic mapping and registration of 3D point clouds. Autom. Constr. **89**, 38–48 (2018). http://www.sciencedirect.com/science/article/pii/S0926580517303990

11. Kim, P., Chen, J., Kim, J., Cho, Y.K.: SLAM-driven intelligent autonomous mobile robot navigation for construction applications. In: Smith, I.F.C., Domer, B. (eds.) Advanced Computing Strategies for Engineering, pp. 254–269. Springer, Cham (2018)

12. Konrad, J., Wang, M., Ishwar, P.: 2D-to-3D image conversion by learning depth from examples. In: 2012 IEEE Computer Society Conference on Computer Vision and Pattern Recognition Workshops, pp. 16–22 (2012)

13. Laina, I., Rupprecht, C., Belagiannis, V., Tombari, F., Navab, N.: Deeper depth prediction with fully convolutional residual networks. CoRR abs/1606.00373 (2016). http://arxiv.org/abs/1606.00373

14. Li, B., Shen, C., Dai, Y., van den Hengel, A., He, M.: Depth and surface normal estimation from monocular images using regression on deep features and hierarchical CRFs. In: IEEE Conference on Computer Vision and Pattern Recognition (CVPR 2015) (2015)

15. Li, B., Dai, Y., Chen, H., He, M.: Single image depth estimation by dilated deep residual convolutional neural network and soft-weight-sum inference. CoRR arxiv:abs/1705.00534 (2017)

16. Li, J., Klein, R., Yao, A.: Learning fine-scaled depth maps from single RGB images. CoRR abs/1607.00730 (2016). http://arxiv.org/abs/abs/1607.00730

17. Liu, B., Gould, S., Koller, D.: Single image depth estimation from predicted semantic labels. In: Proceedings of the Conference on Computer Vision and Pattern Recognition (CVPR) (2010)

18. Liu, F., Shen, C., Lin, G.: Deep convolutional neural fields for depth estimation from a single image. In: Proceedings IEEE Conference on Computer Vision and Pattern Recognition (2015). http://arxiv.org/abs/1411.6387

19. Liu, M., Salzmann, M., He, X.: Discrete-continuous depth estimation from a single image. In: Proceedings of the 2014 IEEE Conference on Computer Vision and Pattern Recognition, CVPR 2014, pp. 716–723. IEEE Computer Society, Washington, DC (2014). https://doi.org/10.1109/CVPR.2014.97
20. Mallick, T., Das, P.P., Majumdar, A.K.: Characterizations of noise in kinect depth images: a review. IEEE Sens. J. **14**(6), 1731–1740 (2014)
21. Nathan Silberman, Derek Hoiem, P.K., Fergus, R.: Indoor segmentation and support inference from RGBD images. In: ECCV (2012)
22. Nguyen, C.V., Izadi, S., Lovell, D.: Modeling kinect sensor noise for improved 3D reconstruction and tracking. In: 2012 Second International Conference on 3D Imaging, Modeling, Processing, Visualization Transmission, pp. 524–530, October 2012
23. Roy, A., Todorovic, S.: Monocular depth estimation using neural regression forest. In: 2016 IEEE Conference on Computer Vision and Pattern Recognition (CVPR), pp. 5506–5514 (2016)
24. Saxena, A., Chung, S.H., Ng, A.Y.: Learning depth from single monocular images. In: Weiss, Y., Schölkopf, B., Platt, J.C. (eds.) Advances in Neural Information Processing Systems 18, pp. 1161–1168. MIT Press (2006). http://papers.nips.cc/paper/2921-learning-depth-from-single-monocular-images.pdf
25. Silberman, N., Hoiem, D., Kohli, P., Fergus, R.: Indoor segmentation and support inference from RGBD images. In: Proceedings of the 12th European Conference on Computer Vision - Volume Part V, ECCV 2012, pp. 746–760. Springer, Heidelberg (2012). https://doi.org/10.1007/978-3-642-33715-4_54
26. Wang, P., Shen, X., Lin, Z., Cohen, S., Price, B., Yuille, A.: Towards unified depth and semantic prediction from a single image. In: 2015 IEEE Conference on Computer Vision and Pattern Recognition (CVPR), pp. 2800–2809, June 2015. https://doi.org/10.1109/CVPR.2015.7298897
27. Yu, Y., Song, Y., Zhang, Y., Wen, S.: A shadow repair approach for kinect depth maps. In: Lee, K.M., Matsushita, Y., Rehg, J.M., Hu, Z. (eds.) Computer Vision - ACCV 2012, pp. 615–626. Springer, Heidelberg (2013)

Person Following Robot Behavior Using Deep Learning

Ignacio Condés[(✉)] and José María Cañas

Universidad Rey Juan Carlos, Móstoles, Spain
ignacio.condes.m@gmail.com, jmplaza@gsyc.urjc.es

Abstract. Human-robot interaction (HRI) is a field with growing impact as robot applications are entering into homes, supermarkets and general human environments. Person following is an interesting capability in HRI. This paper presents a new system for a robust person following behavior inside a robot. Its perception module addresses the person detection on images using a pretrained TensorFlow SSD *Convolutional Neural Network* which provides *robustness* even on tough lighting conditions. It also includes a face detector and a *FaceNet* CNN to reidentify the target person. Care has been put to allow real-time operation. The control module implements two *PID* controllers for a reactive smooth response, moving the robot towards the target person without distracting with other people around. The entire system has been experimentally validated on a real TurtleBot2 robot, with an Asus Xtion RGBD camera.

Keywords: Computer vision · Neural networks · Deep learning
Robotic behavior

1 Introduction

In the last years robots have become a powerful tool for humans in many areas, helping to perform all kind of tasks: hazardous explorations, personal assistance, cleaning, driving, etc. Electronics assembly, car factories, autonomous driving (like Tesla), vaccum cleaners (like Roomba) and even logistics (like Amazon warehouses) are good examples of it. In all these fields robots are designed to perform these tasks on the most autonomous possible way, which means not requiring to be directly controlled on every action. This involves to provide robots a certain intelligence and capabilities to correctly trigger the most suitable action for each possible input stimulus.

As robots enter into offices, supermarkets or homes (like vacuum cleaners, domestic assistants, etc.), better mechanisms for human-robot interaction are also desirable. One useful capability there is the robot being able to follow a particular person.

In addition, one of the main robot sensors are cameras. They are cheap and a very powerful source of information about the robot environment. One of the most popular advances in computer vision are the deep learning techniques,

© Springer Nature Switzerland AG 2019
R. Fuentetaja Pizán et al. (Eds.): WAF 2018, AISC 855, pp. 147–161, 2019.
https://doi.org/10.1007/978-3-319-99885-5_11

using neural networks to process camera images. They can solve classification, detection and segmentation tasks in a very robust way. Typically, neural networks have to be trained on huge datasets and they work fine with new unseen images.

In this paper, a new system for *person following behavior* inside a robot is presented. It avoids the use of laser sensors and relies on a cheap RGBD sensor and color images processing. The person *detection* subsystem uses deep learning to detect people and to identify the particular person the robot has to follow, in particular *Convolutional Neural Networks* (CNNs).

Section 2 reviews some related works about person following and image processing with neural networks. Section 3 describes the design of the proposed system, whose main modules are detailed in the Sects. 4 and 5, explaining its perception side and its control side. It has been implemented and experimentally validated, as it will be seen on Sect. 6. Finally some brief conclusions finish the paper.

2   Related Works

The person following behavior is a old problem in Human-Robot-Interaction field with many existing solutions in the literature [4, 16, 18].

A nice approach [3] uses laser sensors to detect the legs, sensor fusion and a particle filter to provide robustness and continuity to the person estimation.

The topic has been extensively explored using a single camera [16]. Other interesting approaches use stereo vision. [2] combines color of the clothes and position information to detect and track multiple people around. They use Kalman filters and fuse the plain-view map information with a face detector. [17] use depth templates of person shape applied to a dense depth image and an SVM-based verifier for eliminating false positives. They use the Extended Kalman filter to continuously estimate the person positions. In addition, Histogram of Oriented Gradients are a classical and effective method for human detection [1], which is a relevant ingredient of the person following behavior.

In the last years, the availability of low cost RGBD sensors has opened the door to new approaches using them [14, 15], taking advantage of the depth information. The depth simplifies not only the robustness of the detection itself but also the control of the robot movements, which can be the developed as position based controllers. In [9] deep learning techniques on color and motion cues have been used to successfully detect people on RGBD videos for surveillance applications.

The re-identification of a particular person is a complementary problem to general people detection. There are image-based methods and video-based methods. Recent papers also use deep learning to identify the detected persons [13]. Other methods use features and skeleton keypoints [11]. [10] use image and range data combining color, height and gait features (step length and speed) for a robust identification, even in severe illumination environments.

[12] is an interesting work that uses neural networks to both multiple person detection and a particular person re-identification. His system works in real-time on moderate hardware using a single camera.

3   Design

The proposed system follows the structure represented in the Fig. 1 and it is composed of two modules. The *perception* module is responsible for performing the *vision tasks*. Therefore, it determines, with the maximum possible accuracy, the position of every person found inside the color image. It also discerns which of these persons is the one to be followed, in case she is seen. The *actuation* module reads that output from the perception module and sends suitable movement commands to the robot.

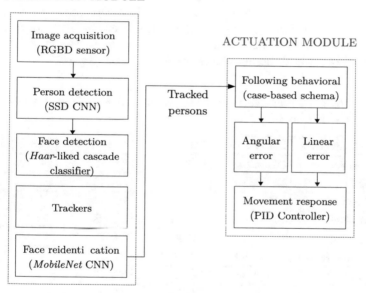

Fig. 1. General design of the proposed system.

These mentioned modules run on a computer (the specifications of our settings can be found at Sect. 6), connected to both the RGBD sensor driver and the robot driver. They have been implemented in Python language as concurrent functions are executed inside a single application. Several threads iterate on independent tasks at an individually defined frequency. In the case of a slow computer the frequency can be lowered down, and so, the system is designed to run properly on different hardware specifications of the base PC.

4   Perception Module

This module is responsible for apprehending the incoming images (RGB and depth) from the sensor, and generates a significant output to be interpreted by the actuation module. Its structure follows a certain pipeline with five steps (Fig. 1): image acquisition, person detection, face detection, person and face trackers, and face reidentification. This pipeline allows the system to discern if the target person is being seen right now.

4.1   Person Detection

First, the incoming RGB images are passed through an *object detection* module. It is powered by a pretrained CNN, which has a special architecture designed in [5] for this purpose: SSD (*Single-Shot Multibox Detector*). This detection model type stands out by its prediction speed, because *it performs a single feed-forward pass* of the image through the network, unlike other state-of-the-art techniques. These other approaches perform successive feed-forward passes through the network, what makes them consequently much slower. Before choosing SSD for the proposed system, several detection architectures were compared and their corresponding typical inference times were measured (Table 1).

Table 1. Timing performance tests for several detection models. The selected implementation is SSD Lite.

Architecture	Base network	Dataset	Mean inference time (ms)
ResNet	Inception	COCO	820.71
SSD	MobileNet	COCO	107.43
ResNet	101	COCO	786.49
ResNet	50	COCO	515.28
ResNet	101	COCO	63.97
Faster-RCNN	ImageNet	ILSVRC2014	703.99
Faster-RCNN	Inception	COCO	352.20
ResNet	50	COCO	793.87
SSD	MobileNet	COCO	102.85
ResNet	101	COCO	898.59
Inception	ResNet	OID	792.42
SSD Lite	**MobileNet**	**COCO**	**68.13**
ResNet	101	Kitti	111.29
Inception	ResNet	OID	667.76

On this module, the desired image to input has to be reshaped to 300 × 300 px, which is a typical shape for this type of CNN architecture. As shown on Fig. 2, it extracts a set of activation maps on its *base network*: the feature extractor (*MobileNet* [19] is used in this particular model), and finally position and class inferences are made later, on multiple image scales. Finally, a *Non Maximum Suppressor* is applied, retaining only the most confident detections, which are adjusted to a correct bounding box shape.

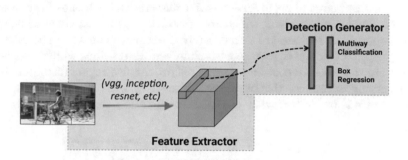

Fig. 2. Schema of the SSD architecture.

This network provides an accurate and efficient object detection, returning for each processed image:

– *Classes:* the detected classes (person, cell phone, airplane, dog, . . .) inferred for each detected object.
– *Scores:* the confidence $\in [0, 1]$ the network has on each object belonging to the estimated class.
– *Boxes:* the coordinates of the rectangular *bounding box* which wraps the detected object, expressed as the coordinates of two opposite corners of it.

Although the used SSD model is capable to detect up to 80 object classes, the proposed system only retains those detection corresponding to *persons*, as it is what we are interested to follow. This whole process provides a lightweight *person detection*, perfectly capable to work in real time on a regular computer.

In addition, our implementation is capable to handle different network models and architectures on a *plug and play* way, just pointing the model file (in the specific TensorFlow .pb format) in the configuration file of the program.

4.2   Face Detection

Once the existing persons on the image have been detected, the next step is to search their faces, in order to know whether any of them corresponds to the target person to follow. In order to achieve it, the classical Viola and Jones face detection algorithm [6] has been used, as it is a simple and fast solution, suitable for a prototype that already handles a *SSD* CNN. This algorithm, which comprises *Haar* features (Fig. 3), is a simple algebraic method which takes advantage of the typical illumination pattern of a face (due to its physical shape) to detect promising regions of the input image to contain a face. To test a *Haar* feature on a specific grayscale image, the feature is slided through it, and a simple operation is performed (black pixels subtracted to white ones). If the result is positive along all the feature, the region where it has been applied passes the test.

As the previous block detected persons inside the image, this face detection algorithm is only applied inside the instance of the detected persons, in order to speed up the system and avoid false positive detections. Each one of the detected persons is divided into *regions*, which are passed through a *cascade* of tests, where the non-compliant regions are immediately discarded. The accepted ones pass to a slightly more complex feature each time, and the regions which pass all the features are supposed to contain a face. This progressive region dismiss makes it an efficient algorithm, capable of run simultaneously with the rest of tasks. This entire process is performed with an OpenCV[1] function.

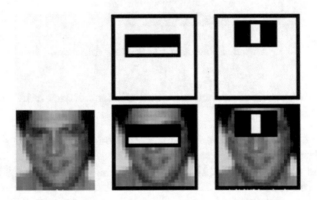

Fig. 3. *Haar* features applied on a face.

4.3   Person and Face Trackers

Both previous steps detect persons and faces, but although they behave in a robust way, their outputs can suffer spurious false negative or false positive detections, due to lighting or occlusions. As they can interfere with the desired following capability, the third step palliates their effect implementing time-spatial *trackers*, which filter the detection outputs (*candidate detections*). The tracker associates to each person/face its coordinates inside the image, and takes into account the number of successive frames on which it has been detected. If it surpasses a *patience* threshold, then it is considered a reliable (*tracked*) detection. Otherwise, it is taken as a spurious output and ignored. In addition, each *tracked detection* is remembered for some frames even without new updates of its current position. This way, a partial occlusion of a person does not cause her elimination, until she has not been seen for a while.

---

[1] Open source image processing library.

This tracking step can be observed in Fig. 4, in a generic *instance* case (which comprises the same behavioral for person and face tracking). In addition, it is not necessary to detect the person/face every time, as the tracker is capable to infer that a new detection near the last position of a person corresponds to its new location, without requiring to see again her face to check whether she is the same person.

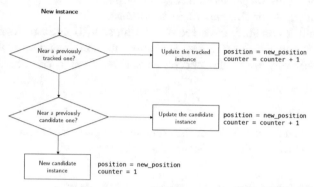

1) For a new instance in the current frame

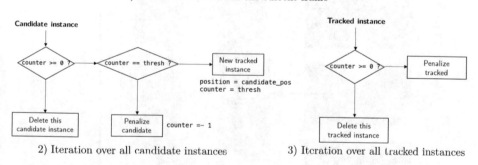

2) Iteration over all candidate instances   3) Iteration over all tracked instances

Fig. 4. Tracking process for a instance (person/face).

4.4   Face Reidentification

The previous trackers turn the unfiltered output of the detection subsystems into tracked and more reliable detections. Hence, they can be used to perform *identification* tasks. For this purpose, a parallel CNN is used, the *FaceNet* [7]. This network *maps* a face image, extracting some key features, into a 128-dimensional Euclidean space, where faces are represented by what is called *embeddings* (feature vectors). The Euclidean L^2 distance (Eq. 1) existing between two of this embeddings stands for the *face similarity* between those faces. Hence, we can consider that two embeddings belong to the same face if their distance is below a threshold, which has experimentally been set to 1.1.

$$d(\boldsymbol{f_1}, \boldsymbol{f_2}) = \sqrt{\sum_{i=1}^{128}(f_{1_i} - f_{2_i})^2} \tag{1}$$

The proposed solution makes use of this, and computes the embeddings of each detected (and tracked) face on real time. Once this has been performed, it compares the similarity between these embeddings and the one corresponding to the target person. This target embedding is computed when the program starts, from a given image file of the *person to be followed*.

To avoid penalties on similarity due to lighting conditions, a previous blurring and *prewhitening* (on Eq. 2, with x as the color channel, μ as its mean and σ as its standard deviation) are performed on each given face to the *FaceNet*.

$$x' = \frac{x - \mu}{\sigma} \tag{2}$$

In Fig. 5 the same face is seen in different lighting conditions, and all of them yield embeddings with a distance lower than the threshold.

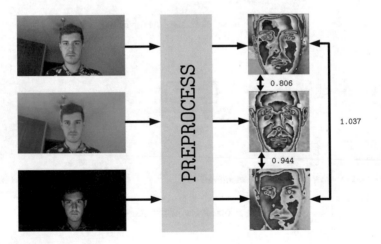

Fig. 5. Relative distances between the same face on different lighting situations.

5   Actuation Module

With the *perception* module the system obtains the maximum possible certainty about the persons present in the current RGB image, as well as their condition of being or not the one to be followed. The *actuation* module is responsible of generating a suitable command to the robot motors, in order to move towards the target person, in case it is being tracked in the current frame. To do so, it follows its own pipeline, explained next.

The input for this module is the information yielded by the *perception* module: the *tracked persons* parameters (position, face, "is or is not the target person"). From this information, it has to infer the proper robot movements taking also into account the state of the system on the last iteration. This is implemented with a *case-based* behavior, which follows the *flow chart* represented on Fig. 6. The response mainly depends on *mom* (the name given to the target person), as it can be *lost* or *tracked and followed.*

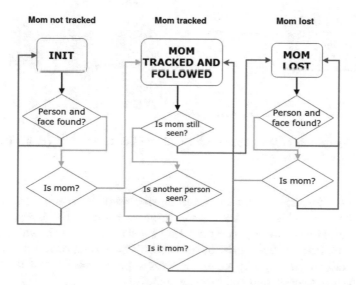

Fig. 6. *Flow chart* followed by the case-based behavioral, depending on the previous state.

If the target person is found among the tracked persons, the system follows it. It can be observed that the system does not need a continuous face feedback to follow the target person. The time-spatial continuity provided by the previously described trackers allows a stable tracking even when the person is only visible from her back side. In addition, if a new face is found and it satisfies the criteria of similarity with the reference face, then the *followed* role is switched to that new person, which starts being followed by the robot.

This actuation module has been implemented within a computer iterative thread, and its pipeline is executed *once* per thread iteration. This results on a new speed command pair (angular and linear speeds) to the motor base on each iteration.

5.1   Error Computation

Once the system has recognized the target person inside the image, in case it is being seen, it proceeds to an *error computation,* in order to determine the strength of the required rotation and translation to move towards that person. As the robot moves over the ground plane, with two degrees of freedom, the system computes two errors: angular error and distance error.

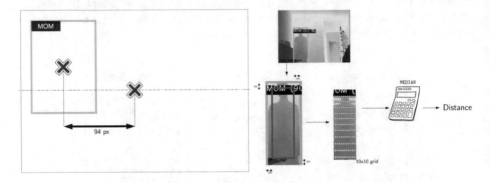

Fig. 7. Computation of the angular error (left) and distance error (right) between the system and the followed person.

The most desirable robot orientation with respect to the person is to be aligned with her. This way, the person appears right in the horizontal center of the image. Hence, we can compute the *angular error* as the subtract of the center coordinates and the center of the person's bounding box, in the horizontal dimension (Fig. 7 (left)). This error gives an approximate idea of the necessary turn to align the robot and the target person.

The used sensor is a RGBD camera, which provides *depth* images. These depth images are aligned with the RGB ones, which means that the coordinates of the person inside the depth image are the same than the RGB ones. Hence, it allows to *locate* the person inside the depth map and measure the *distance error*. For the sake of robustness, this has been implemented using a 10 × 10 grid sampling of the depth values inside the person box (putting care on avoiding the margins, in order to measure only inside the person). This allows to collect a set of measured distances to the person. It computes the *median* of that set of measurements, taking the result as the real distance from the robot to the person, as seen on Fig. 7 (right).

5.2   Movement Control

Both angular and distance errors are computed to determine the *relative position* of the robot and the target person. They are used as an input to compute the most suitable control response. As the purpose is not to move the robot literally to the position of the person, but to just maintain a following behavior, a *desirable zone* is established in each degree of freedom. When the target person is found inside these desirable zones (illustrated on Fig. 8), the robot will not

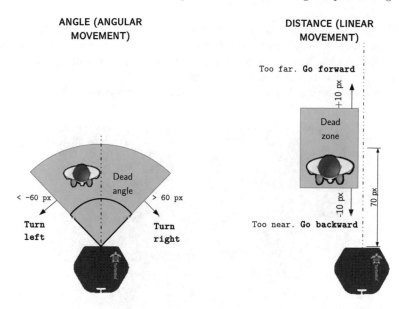

Fig. 8. Dead zones on each dimension, where the person is considered under control.

move towards her, as the person is considered *under control*. They are also known as dead zones as no correction from control is needed.

If the person is outside these dead zones, a physical correction response is required, with the purpose of *seeing* that person again inside the dead zone. For this action, each dimension implements a *PID* controller, which establishes a *closed-loop* feedback, as described on [8]. This allows to keep in mind previous responses, to achieve the optimum fitting on each iteration. This means, for example, to accelerate if the person is not going any closer, or to step hard on the brake if the person suddenly gets too close.

The implemented *PID* controllers (an angular and a linear one) have been experimentally tuned to obtain the most suitable parameters for our operation, obtaining the values in Table 2.

Table 2. Optimal found values for the parameters in each PID controller.

	Linear	Angular
k_p	2	7
k_d	0.1	0.5
k_i	3	10

This way, the system can output a speed command with a tight adjustment to the values required by the situation of the current iteration. The response obtained from this value is a *reactive* one. This means that each value results in a new movement command, avoiding to perform movements longer than one iteration. For smoothness sake, what is sent to the motors is not that value, but the *mean* between it and the last sent one. This way, it results on a slightly longer convergence that helps to remove sudden movements.

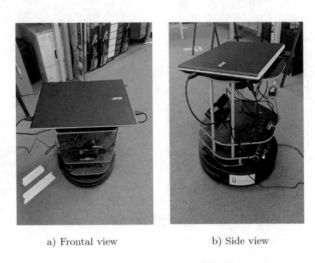

a) Frontal view   b) Side view

Fig. 9. Robot used for the validation experiments.

6   Experiments

The developed system has been experimentally validated on a real robot and an indoor scenario. The TurtleBot-2 robot was used on the experiments, including an Asus Xtion Pro Live RGBD sensor. The onboard computer was a regular laptop, equipped with an Intel i5-4210U CPU, a DDR3 RAM of 8 GB and a Nvidia 940M GPU (Fig. 9). On these hardware resources (using GPU parallelization to run the neural networks, and choosing a lightweight *SSD Lite* model), we achieved a stable detection rate of 14 fps, fairly enough to guide the robot on a seamless way.

The person detection and reidentification process can be observed on Fig. 10. The system is successfully able to re-identify the target person (*mom*), even in challenging lightning conditions like in Fig. 10.

The complete system results show a robot finely following a particular person, even when that person does not face towards the robot. The behavior is fluent, with a refresh rate of 10 movements per second (as the SSD CNN is the lightest possible model, as seen on Table 1). This agility of the neural network for people detection helped substantially to minimize the time bottleneck. The following behavior can be observed on Fig. 11. Several videos of a typical execution are also publicly available[2].

---

[2] Full video test available on https://www.youtube.com/watch?v=oKMR_QCT7EE.

Fig. 10. Detection and face re-identification of the target person.

Fig. 11. Following behavior.

7   Conclusions

In this paper an algorithm for person following behavior inside a robot has been presented. The proposed system has been implemented with two concurrent modules in a Python program, one for *perception* and another one for *actuation*.

For perception it uses a pretrained people detection network, a regular face detector based on Haar features, some filters to alleviate false positives and false negatives, and a FaceNet neural network for person re-identification. The visual information is combined with depth data from the RGBD sensor to estimate the relative position of the target person from the robot. The person detection network was carefully selected to meet the real time operation requirement, typical in robotics. All this leads into a fast and efficient following system, capable of keeping the track on a particular individual even without a continuous checking of her identity or detection of her face.

For actuation it uses two PID controllers which are based on angular error and distance error between the robot and the target detected person.

All these functionalities have been combined into a software node that runs on real time. This allows a simple robot to perform a successful tracking of a person, just knowing her face beforehand. The source code of the entire system is available online[3].

Several experiments with a real robot have been performed and are also publicly available. They validate the successful operation of the proposed system.

---

[3] https://github.com/roboticsurjc-students/2017-tfg-nacho_condes.

The use of deep learning neural networks provide high robustness to the image based person detection, which works even on tough lighting conditions. This has been the main design restriction developing this system, which can serve as a first prototype of a powerful people following platform.

This work can be extended in several lines. For instance using another neural network for face detection instead of Haar features. In addition, the use of faster networks for people detection, like YOLO DarkNet ones, is being explored. For this purpose, a rewriting of the node software in C++ would be required.

References

1. Dalal, N., Triggs, B.: Histograms of oriented gradients for human detection. In: IEEE Computer Society Conference on Computer Vision and Pattern Recognition CVPR (2005)
2. Muñoz-Salinas, R., Aguirre, E., García-Silvente, M.: People detection and tracking using stereo vision and color. Image Vis. Comput. **25**(6) (2007)
3. Aguirre, E., García-Silvente, M., Pascual, D.: A multisensor based approach using supervised learning and particle filtering for people detection and tracking. In: Robot 2015: Second Iberian Robotics Conference. Springer (2016)
4. Calvo, R., Cañas, J.M., Garcíía-Pérez, L.: Person following behavior generated with JDE schema hierarchy. In: Poster in ICINCO 2nd International Conference on Informatics in Control, Automation and Robotics, pp. 463–466. INSTICC Press, Barcelona (Spain), 14–17 September 2005. ISBN: 972-8865-30-9
5. Lui, W., Anguelov, D., Erhan, D., et al.: SSD: Single-Shot Multibox Detector. CoRR, abs/1512.02325 (2015)
6. Viola, P., Jones, M.: Rapid object detection using a boosted cascade of simple features. In: Proceedings of the 2001 IEEE Computer Society Conference on Computer Vision and Pattern Recognition CVPR, vol. 1, pp. I-511–I-518 (2001)
7. Schroff, F., Kalenichenko, D., Philbin, J.: FaceNet: A unified embedding for face recognition and clustering. CoRR, abs/1503.038032 (2015)
8. Åström, K.J., Murray, R.M.: Feedback Systems: An Introduction for Scientists and Engineers (2004)
9. Xue, H., Liu, Y., Cai, D., He, X.: Tracking people in RGBD videos using deep learning and motion clues. Neurocomputing **204**, 70–76 (2016)
10. Koide, K., Miura, J.: Identification of a specific person using color, height, and gait features for a person following robot. Robot. Auton. Syst. **84**, 76–87 (2016)
11. Munaro, M., Ghidoni, S., Tartaro, D.D., Emanuele, M.: A feature-based approach to people re-identification using skeleton keypoints. In: 2014 IEEE International Conference on Robotics and Automation (ICRA)
12. Welsh, J.B.: Real-Time Pose Based Human Detection and Re-identification with a Single Camera for Robot Person Following. Ph.D. Thesis, University of Maryland, College Park (2017)
13. Yoon, Y., Yoon, H., Kim, J.: Person reidentification in a person-following robot. In: 25th IEEE International Symposium on Robot and Human Interactive Communication (RO-MAN) (2016)
14. Shimura, K., Ando, Y., Yoshimi, T., Mizukawa, M.: Research on person following system based on RGB-D features by autonomous robot with multi-kinect sensor. In: 2014 IEEE/SICE International Symposium on System Integration (SII)

15. Ilias, B., Shukor, S.A.A., Yaacob, S., Adom, A.H., Razali, M.H.M.: A nurse following robot with high speed Kinect sensor. ARPN J. Eng. Appl. Sci. **9**(12), pp. 2454–2459 (2014)
16. Yoshimi, T., Nishiyama, M., Sonoura, T., Nakamoto, H., Tokura, S., Sato, H., Ozaki, F., Matsuhira, N., Mizoguchi, H.: Development of a person following robot with vision based target detection. In: 2006 IEEE/RSJ International Conference on Intelligent Robots and Systems
17. Satake, J., Miura, J.: Robust stereo-based person detection and tracking for a person following robot. In: ICRA Workshop on People Detection and Tracking (2009)
18. Sidenbladh, H., Kragic, D., Christensen, H.I.: A person following behaviour for a mobile robot. In: Proceedings of the IEEE International Conference on Robotics and Automation (1999)
19. Howard, A.G., Monglong, Z., et al.: MobileNets: efficient convolutional neural networks for mobile vision. CoRR J. **1704.04861** (2017)

Human Robot Interaction

Attentional Mechanism Based on a Microphone Array for Embedded Devices and a Single Camera

Antonio Martinez-Colon$^{(\boxtimes)}$, Jose M. Perez-Lorenzo$^{(\boxtimes)}$, Fernando Rivas$^{(\boxtimes)}$, Raquel Viciana-Abad$^{(\boxtimes)}$, and Pedro Reche-Lopez$^{(\boxtimes)}$

Multimedia and Multimodal Processing Research Group, Telecommunication Engineering Department, University of Jaén, Linares, Spain
{amcolon,jmperez,rivas,rviciana,pjreche}@ujaen.es

Abstract. This work presents an attentional mechanism with the capability of detecting the localization of a speaker for interaction purposes, based on audio and video information. The localization is computed in terms of azimuth and elevation angles, to be used as input values for controlling mobile systems such as a pan-tilt videocamera or a robotic head. For this purpose the SRP-PHAT algorithm has been implemented with a commercial array of microphones for embedded devices, in order to estimate the localization of a sound source in the surroundings of the array. In order to improve the limitations of the SRP-PHAT algorithm in the estimation of the z coordinate, the elevation angle is corrected via video information by using Haar cascade classifiers for face detection. Simulations and experiments show the accuracy of the system, as well as the application for controlling a pan-tilt videocamera in a real scenario with speakers and ambient noise.

Keywords: Attentional mechanism · Audio source localization
Microphone array · Face detector

1 Introduction

In order to obtain an effective interaction between humans and machines or physical agents, the machines have to be able to localize people in its surroundings [1]. Moreover, machines have to pay attention to the human voice and look at the possible speakers. In a dynamical situation where the speakers are in movement the physical agents should be able to track them. The development of a good attentional mechanism has many applications in a great variety of engineering and research fields such as social robotics, video-conference systems and speaker diarization for meetings.

In this paper, an attentional mechanism has been developed based on audio and visual information. These data are used in order to localize possible human speakers. In our approach an acoustical azimuthal localization is performed by

© Springer Nature Switzerland AG 2019
R. Fuentetaja Pizán et al. (Eds.): WAF 2018, AISC 855, pp. 165–178, 2019.
https://doi.org/10.1007/978-3-319-99885-5_12

means of SRP-PHAT algorithm [2] whereas the visual localization is performed with a camera using the OpenCV Haar-cascade face and eyes detector [3] to set the elevation angle. In our proposal an array with only 4 microphones geometrically closed among them is used. The maximum separation among microphones is 8.6 cm which allows its utilization in compact systems of videoconference as well as in social robotics.

Related work on person tracking in the field of social robots includes [4–7]. In [4] an array of 8 microphones mounted on a rectangular prism of dimensions 50 cm × 40 cm × 36 cm is used for sound source localization on a mobile robot. Nakamura et al. use an extension of MUSIC algorithm in a robot-embedded 8-channel circular microphone array, along with thermal and distance cameras [5]. In contrast to these works, our proposal is based on an array with a lower number of microphones and with a smaller size. Also, binaural architectures are found in the literature. As examples, Ferreira et al. propose a framework for perception by fusing vision, vestibular sensing and binaural auditory system [6]. Viciana-Abad et al. make a bioinspired proposal using a pair of microphones and a stereo vision camera system [7]. However, in binaural approaches, the audio system may present problems in the accuracy of the detection in noisy environments and in the front-back ambiguity. In our work, the use of a planar array instead of a linear one allows to detect the direction of arrival of any speaker 360° around the array without ambiguity. Respect to the applications for videoconference systems and diarization, most of them use in general microphones that are distributed within the room and very far away each others [8–10], frequently with a great number of microphones [11].

The rest of this paper is organized as follows. Section 2 presents the fundamentals of SRP-PHAT, follows with the implementation of the algorithm, and ends with simulations with a MATLAB software. Next, Sect. 3 describes briefly the visual features that are included in the system. Section 4 presents the results in the audio localization task with two different arrays, being one of them a commercial one for embbeded devices, and comparing the implemented algorithm also with a commercial one. At the end of the Section some results with the whole system are also shown. Finally, conclusions and future work are described in Sect. 5.

2   Audio Source Localization

2.1   Introduction to SRP-PHAT

An array of microphones can be defined as a set of N microphones, where every microphone is located at an unique position. Then, in a simple model, it can be supposed that the acoustic waves originated at a sound source follow a direct path to every microphone along N lines simultaneously. The orientation of these lines defines the direction of the propagation vectors (Fig. 1) and the time of arrival of the waves at every microphone will depend on these vectors. Methods for the localizacion of acoustical sources using an array of microphones can be divided into two groups: indirect and direct. The indirect methods are typically

based on two steps. First, the *Time Difference of Arrival* (TDOA) is estimated
for each of the existing pairs of microphones within the array. Then, based on
these time delay estimation (TDE) values, the localization of the acoustic source
can be estimated if the geometry of the microphones array is well known. On the
other hand, the direct methods are able to compute both TDOA and localization
estimations in a single step. For that purpose, direct methods perform a scanning
of localizations in the surroundings of the microphones array, and the most
likely position is selected as the estimation of the acoustical source. Both type
of approaches, indirect and direct, are also often based on the General Cross
Correlation (GCC) technique [12]. If an array of microphones is used, the GCC
for two signals captured by a pair of microphones (k, l) is defined as:

$$R_{m_k m_l}(\tau) = \int_{-\infty}^{\infty} \psi_{kl}(w) M_k(w) M_l^*(w) e^{jw\tau} dw \qquad (1)$$

being τ a time delay, $M_k(w)$ and $M_l(w)$ are the Fourier Transform of the microphone signals $m_k(t)$ and $m_l(t)$ respectively, $*$ denotes the complex conjuga-
tion, and $\psi_{kl}(w)$ is known as the generalized frequency weighting. The choice of
this weighting function may influence on the performance of the method under
adverse acoustic conditions [13], and the Phase Transform (PHAT) weighting is
one of the most widely used, which is defined as:

$$\psi_{kl}(w) = \frac{1}{|M_k(w) M_l^*(w)|} \qquad (2)$$

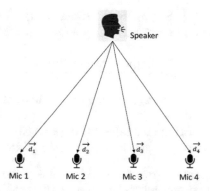

Fig. 1. Propagation vectors for an array of 4 microphones

One of the best known direct algorithms for the localization of sound sources
is the Steered Response Power-Phase Transform (SRP-PHAT) [14,15], which
uses the PHAT weighting for the computation of the GCC for every pair of microphones inside the array. The SRP-PHAT algorithm is commonly interpreted as
a beamforming method which is used to distinguish the spatial properties of a

target signal from a background noise. For this purpose, the position of a candidate source that maximizes the output of a steered delay-and-sum beamformer is searched. The SRP-PHAT algorithm has been proved to be a robust method under adverse conditions, although its computational cost may be a problem for real time applications.

2.2   Implementation

In this subsection the implementation made for the SRP-PHAT algorithm is described. The steered response power (SRP) at the surroundings of the array can be expressed as:

$$P_n(\mathbf{x}) = \int\limits_{nT}^{(n+1)T} |\sum_{l=1}^{N} w_l m_l(t - \tau(\mathbf{x}, l))|^2 dt \qquad (3)$$

being $\mathbf{x}$ the spatial point where the SRP is computed, n the time frame of length T, N the number of microphones, $m_l(t)$ denotes the signal output for a given microphone l, w_l is a weight, and $\tau(\mathbf{x}, l)$ is the propagation time of the direct path from the point $\mathbf{x}$ towards the microphone l.

Removing some terms of fixed energy, the part of $P_n(\mathbf{x})$ that is variable with $\mathbf{x}$ can be expressed as [2]:

$$P'_n(\mathbf{x}) = \sum_{k=1}^{N} \sum_{l=k+1}^{N} R_{m_k m_l}(\tau_{kl}(\mathbf{x})) \qquad (4)$$

being $R_{m_k m_l}$ the GCC for the pair of microphones (k, l) as defined in (1), and $\tau_{kl}(\mathbf{x})$ is the Inter-Microphone Time-Delay Function (IMTDF), which can be expressed as

$$\tau_{kl}(\mathbf{x}) = \frac{||\mathbf{x} - x_k|| - ||\mathbf{x} - x_l||}{c} \qquad (5)$$

being c the speed of sound, and x_k, x_l the points where microphones k and l are respectively located.

The SRP-PHAT algorithm evaluates (4) over a grid of the space in order to find a maxima, and typically it can be implemented following next steps:

1. Grid Definition. A spatial grid is defined with a given resolution. Then, the theoretical delays between all the points of the grid and every pair of the microphones that can be formed are precomputed. This is done only once at the beginning of the application, and these precomputed delays depend on the geometry of the array. In this work, the size of the grid cells has been set to 0.1 m.
2. GCC computation. For the selected audio frame, the GCC for every pair of microphones is computed following (1).
3. For every point in the grid, the cross-correlation values are accumulated following (4).

4. Selection of the source position. The location **x** of the active source is selected as the point of the grid with a maximum value.

The cost computation in the implementation of the algorithm is a key aspect in applications that must interact with a user. Different types of optimization exist in the literature, and in this work a set of equispaced delays is used in step 2 and 3. By this way, instead of computing the GCC for the delays resulting from every combination of grid points and pairs of microphones, it is computed for a set of equispaced delays, with the limit values depending on the size of the array:

$$-\frac{d}{c} \leq \tau \leq \frac{d}{c} \tag{6}$$

being d the diameter of the array. Then, when accumulating the values of cross-correlations in (4), for the estimation of (1) given a delay τ_{kl}, the value is computed as the linear interpolation of the cross-correlation values corresponding to the two closest delays from the set. From now on, the number of equispaced delays will be denoted as n_τ, and its selection has been made based on a set of simulations as explained in SubSect. 2.3. For the application of the algorithm given a stream of audio, Hann windows of 25 ms length have been used over frames of 2 s (see Fig. 2). The SRP-PHAT is computed for every window and accumulated through the frame, and the position of the source is selected as the mode of the accumulated values.

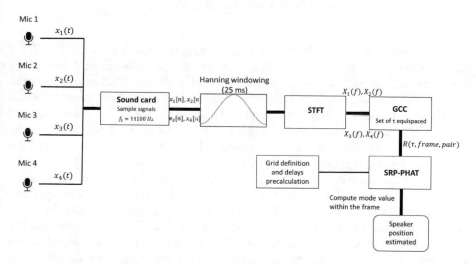

Fig. 2. SRP-PHAT applied to frames of 2 s length with windows of 25 ms.

2.3   MATLAB Simulations

The simulation software RoomSim for MATLAB [16] has been used to decide the initial configuration of the system. This simulator is a useful tool since it allows to generate the impulse response between an audio source and a microphone in a simulated room with a shoe-box shape (Fig. 3). The simulator allows to locate both the source and the microphone in any coordinate (x, y, z) of the space. Therefore, a simulation of an audio signal captured by every microphone of an array can be obtained by modifying the microphone coordinates through the different positions within the microphone array (Fig. 1), and afterwards computing the convolution of every obtained impulse response with an anechoic audio signal.

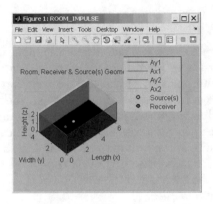

Fig. 3. RoomSim simulator

Based on these simulations, different configurations for the SRP-PHAT implementation have been tested. Since the number of equiespaced delays affect the computation cost of the algorithm, sets with different values for n_τ have been tested: 181, 362 and 543. Simulations have shown that increasing the number above 181 does not improve significantly the accuracy on the detection, and however the computational cost is incremented. Some of the simulations results regarding the detection accuracy are shown in Table 1. The selected configuration has been a squared array of 4 microphones located in the plane $z = 0$ and with a distance between the microphones of 0.3 m. On the other hand, a voice source has been located at the coordinates (4.0, 4.5, 1.7) m. The grid has been limited to a size of $(5.5 \ x \ 9.0 \ x \ 3.3)$ m with a resolution of 0.1 m for every dimension. It can be seen that the accuracy in the detection is the same for the different sets with a value of n_τ above 181. Also, the simulations show that the accuracy for x and y coordinates is high, but on the contrary, the accuracy in the z coordinate is quite poor. In consequence, the estimation for the azimuth angle, which depends on x and y coordinates, is suitable for directing the attentional mechanism, but for the elevation angle, which depends on y and z coordinates, an alternative way to improve the estimation must be found.

Table 1. Accuracy in the estimation of a simulated speaker at (4.0, 4.5, 1.7) m.

n_τ	x	y	z
181	3.9	4.7	0.2
362	3.9	4.7	0.2
543	3.9	4.7	0.2

3   Face Detection

In order to overcome the limitation on the estimation for the elevation angle, the integration of visual information has been used. Since the purpose of the application is to direct the attention towards people speaking, a face detection is applied over the video streaming captured by the camera (Fig. 4). While the control of the azimuth position is done mainly based on the audio information, the visual information is used in order to control the elevation angle, as well to adjust the azimuth value in order to center on the image the face of the person who is detected. For this purpose, Haar cascade classifiers [3] for the detection of both frontal faces and eyes features are applied with OpenCV library.

Fig. 4. Face detection with OpenCV library

4   Results

Initially, an array of four omnidirectional AKG C 417 PP microphones with the same geometry as the one simulated in Subsect. 2.3 has been used in a real environment. An experiment has been carried out to test the accuracy in the detection of the azimuth angle where a person has been speaking in different positions respect to the array, at an average distance of 1.75 m. Also, for a more realistic situation, typical noisy sound sources such as an air conditioner or computer keyboards were active in the room. In order to establish the ground truth for the azimuth values, a laser based angle measurer has been used. While the parameters of the SRP-PHAT implementation have been the same as in the simulations, additionaly a configuration with a lower value for n_τ has been tested for a lower response time of the attentional mechanism. Table 2 summarizes the results in the estimation for the speaker position, while he/she has been moving around the array. It can be seen that for $n_\tau = 80$ the accuracy of the results is slightly lower, but still valid for the purpose of localizing the speaker, since the error is at most of $2.7°$.

Table 2. Accuracy in the estimation of the azimuth angle with a squared array of 0.3 m length. The values represent the average of three measures for each position. The speaker was located at an average distance of 1.75 m from the microphone array.

Ground truth	Squared array, $n_\tau = 181$	Squared array, $n_\tau = 80$
0°	0°	0°
30°	29.5°	28.5°
60°	60.5°	59.0°
90°	90.0°	90.0°
120°	118.8°	118.6°
150°	149.2°	149.0°
180°	180.0°	180.0°
210°	208.4°	208.0°
240°	238.4°	237.3°
270°	270.0°	270.0°
300°	299.3°	298.3°
330°	329.3°	329.3°

At a second stage, the algorithm has been implemented with a commercial microphone circular array for embedded devices from XMOS Company (see Fig. 5), that enables developers and equipment manufacturers to add far-field voice capture to consumer electronics and IoT products, and has been qualified by Amazon for the Alexa Voice Service [17]. The array incorporates 7 PDM microphones, 6 of them forming a circle with a diameter of 8.6 cm, and 1 microphone is located at the center. While the device driver allows to capture the raw audio from all the microphones via USB, for this work only the audio signal captured by 4 of them have been used, with the positions depicted in Table 3. In the implementation of this system, for keeping a lower computational cost, the value for n_τ has been kept to 80. Besides, another XMOS device (xCORE Vocal-Fusion Speaker kit) that incorporates an on-board computation of the azimuth angle of speaker sources has been used for comparison purposes, since it is based on a similar circular array. By this way, the results of the azimuth detection algorithm has been compared with a state-of-the-art commercial product with the same hardware.

A similar evaluation to the previous one has been carried out for these setups, being the results shown in Table 4. It can bee seen that the SRP-PHAT based algorithm is more accurate in the localization task than the on-board built Vocal-Fusion one. Figure 6 depicts the error of the estimations respect to the ground truth.

(a)   (b)

Fig. 5. (a) Commercial circular array of 7 PDM microphones from XMOS Company (b) Experiments setup

Table 3. (x,y) coordinates of the locations of the 4 microphones used within the XMOS circular array. The origin (0,0) is the center of the array.

	$x(cm)$	$y(cm)$
Microphone #1	3.75	2.15
Microphone #2	−3.75	2.15
Microphone #3	3.75	−2.15
Microphone #4	−3.75	−2.15

Table 4. Accuracy in the estimation of the azimuth angle. The values represent the average of three measures for each position. The speaker was located at an average distance of 1.75 m from the microphone array.

Ground truth	VocalFusion	SRP-PHAT
0°	0°	0°
30°	31.7°	30.8°
60°	62.3°	60.1°
90°	94.3°	90°
120°	121.3°	120.5°
150°	143.7°	150.1°
180°	180°	180°
210°	215°	209.5°
240°	240.7°	240.3°
270°	264°	270°
300°	292.7°	299.7°
330°	323°	329.7°

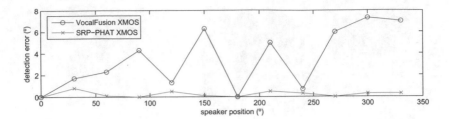

Fig. 6. Absolute value of the azimuth estimation error vs speaker position. The values represent the average of three measures for each position.

Despite the accuracy in the estimation of the azimuth angle, the estimation in the z coordinates was poor in the experiments, and thus the elevation angle is not valid for the attentional mechanism, similarly to the results of the previous simulations. This fact makes necessary to use an alternative way to estimate these values, which in this work it has been done through the analysis of video frames, searching for the faces of the speakers. Respect to the video setup, a Pan-Tilt-Zoom network AXIS videocamera has been used (Fig. 7), which can be controlled via HTTP and incorporates a video streaming server.

Fig. 7. AXIS V5914 videocamera

Several experiments in a live situation have been carried out, where it has been tested that the camera is able to aim to the person that is speaking in a room with several speakers. The pan movement is autonomously controlled based on the audio localization, while the face detection is used for an adjustment of both pan and tilt angles in order to center the speaker face in the video frames (Fig. 8). To avoid a continuous readjustment of the camera trying to center the face, a threshold value has been set as a displacement limit for the coordinates of consecutive detected faces before updating the camera. This value has been set experimentally to 20 pixels for a comfortable visualization at the client side of the video streaming. Also, the average computation time for the SRP-PHAT audio localization takes around 1s using a laptop with an i7 processor. Some relevant captures of the experiments are depicted through Figs. 9, 10 and 11. In Fig. 9a it can be seen a frame where the camera is pointing towards an active speaker in the room. Then, a second speaker begins to speak in Fig. 9b and the camera changes its pointing orientation to this second speaker. Figure 10a–d try

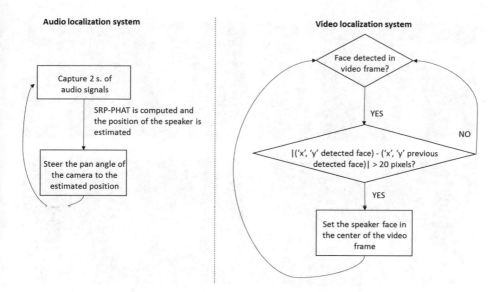

Fig. 8. Pan-Tilt Control of the videocamera

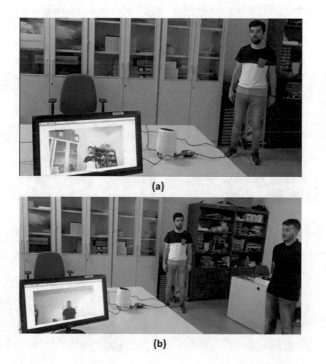

(a)

(b)

Fig. 9. Experiments. The monitor in the image allows to see the video sent from the camera server to a HTTP client. (a) There is one person speaking in the room, and the camera is pointing towards the speaker (b) A second person starts to speak, and the camera updates its orientation.

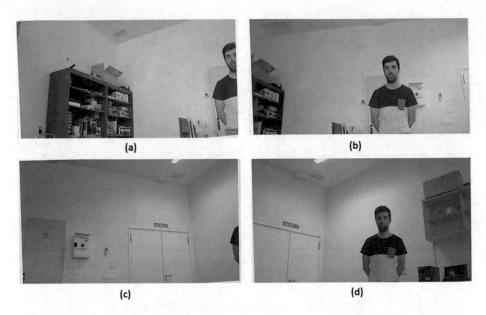

Fig. 10. Experiments. The figures depict the images of the monitor screen (a) Speaker has moved towards his left (b) The camera changes its orientation (c) Speaker moves again (d) The camera changes again its orientation

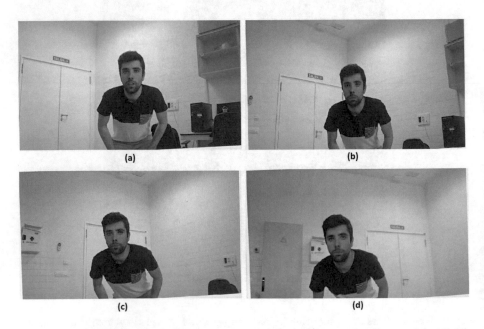

Fig. 11. Experiments. The figures depict the images of the monitor screen (a)–(d) The speaker is moving towards his right side, and the camera corrects its orientation in order to center his face

to show the behaviour of the system when a person moves in the room while speaking. In this case, Fig. 10a and c show a movement of the speaker towards his left side, and Fig. 10b and d show how the videocamera moves towards the speaker, even when the speaker is outside the field of view, since the localization is based on the audio information. Finally, Fig. 11a–d show a situation where the camera follows the speaker based on the face detection, where the system is trying to get centered the detected face on the screen.

5   Conclusions and Future Work

In this work an attentional mechanism based on a microphone array for embedded devices has been presented. While SRP-PHAT is a widely used technique with arrays of microphones, its application has been barely reported in arrays of reduced dimensions and with a low number of microphones, being the main contribution of this work. It has been shown that, while it can achieve a high accuracy in the x and y coordinates for estimating the azimuth angle of a sound source, the z coordinate fails in this estimation. To overcome this limitation, a video based face detection is proposed to correct the elevation angle. The experiments have shown the viability of the system in the interaction with people speaking while moving around. Further work will be focused on different items, such as the implementation of a more robust control algorithm based on probabilistic fusion of audiovisual information, the recognition of the type of audio sources in the environment, the incorporation of a zoom functionality to the pan/tilt movement, the application within a hardware for a robotic head, or the integration of the algorithm in a robotic framework such as Robocomp [18].

Acknowledgements. This work has been supported by Economy and Competitiveness Department of the Spanish Government and European Regional Development Fund under the project TIN2015-65686-C5-2-R (MINECO/FEDER, UE).

References

1. Fong, T., Nourbakhsh, I., Dautenhahn, K.: A survey of socially interactive robots. Robot. Auton. Syst. **42**(3), 143–166 (2003)
2. DiBiase, J.: A high-accuracy, low-latency technique for talker localization in reverberant environments using microphone arrays. Ph.D. Thesis. Brown University (2000)
3. Viola, P., Jones, M.: Rapid object detection using a boosted cascade of simple features. In: Proceedings of the 2001 IEEE Computer Society Conference on Computer Vision and Pattern Recognition CVPR, vol. 1, pp. 511–518 (2001)
4. Valin, J.M., Michaud, F., Rouat, J., Letourneau, D.: Robust sound source localization using a microphone array on a mobile robot. In: International Conference on Intelligent Robots and Systems (IROS 2003), vol. 2, pp. 1228–1233 (2003)
5. Nakamura, K., Nakadai, K., Asano, F., Ince, G.: Intelligent sound source localization and its application to multimodal human tracking. In: IEEE/RSJ International Conference on Intelligent Robots and Systems, pp. 143–148 (2011)

6. Ferreira, J., Lobo, J., Bessiere, P., Castelo-Branco, M., Dias, J.: A Bayesian framework for active artificial perception. IEEE Trans. Cybern. **43**(2), 699–711 (2013)
7. Viciana-Abad, R., Marfil, R., Perez-Lorenzo, J.M., Bandera, J.P., Romero-Garces, A., Reche-Lopez, P.: Audio-visual perception system for a humanoid robotic head. Sensors **14**(6), 9522–9545 (2014)
8. Do, H., Silverman, H.F., Yu, Y.: A real-time SRP-PHAT source location implementation using stochastic region contraction (SRC) on a large-aperture microphone array. In: 2007 IEEE International Conference on Acoustics, Speech and Signal Processing - ICASSP 2007, vol. 1, pp. 121–124 (2007)
9. Do, H., Silverman, H.F.: A fast microphone array SRP-PHAT source location implementation using coarse-to-fine region contraction (CFRC). In: 2007 IEEE Workshop on Applications of Signal Processing to Audio and Acoustics, pp. 295–298 (2007)
10. Marti, A., Cobos M., Lopez, J.J.: Real time speaker localization and detection system for camera steering in multiparticipant videoconferencing environments. In: IEEE International Conference on Acoustics, Speech and Signal Processing (ICASSP), pp. 2592–2595 (2011)
11. Silverman, H., Yu, Y., Sachar, J., Patterson, W.: Performance of real-time source-location estimators for a large-aperture microphone array. IEEE Trans. Speech Audio Process. **13**(4), 593–606 (2005)
12. Knapp, C., Carter, G.: The generalized correlation method for estimation of time delay. IEEE Trans. Acoust. Speech Signal Process. **24**(4), 320–327 (1976)
13. Perez-Lorenzo, J.M., Viciana-Abad, R., Reche-Lopez, P., Rivas, F., Escolano, J.: Evaluation of generalized cross-correlation methods for direction of arrival estimation using two microphones in real environments. Appl. Acoust. **73**(8), 698–712 (2012)
14. DiBiase, J.H., Silverman, H.F., Brandstein, M.S.: Microphone arrays: signal processing techniques and applications. In: Brandstein, M.S., Ward, D. (Eds.) Springer (2001)
15. Marti, A.: Multichannel audio processing for speaker localization, separation and enhancement. Ph.D. Thesis. Universitat Politècnica de València (2013)
16. Campbell, D.R., Palomäki, K.J., Brown, G.J.: A MAT-LAB simulation of shoebox room acoustics for use in re-search and teaching. Comput. Inf. Syst. J. **9** (2005)
17. http://www.xmos.com/
18. Manso, L.J., Bachiller, P., Bustos, P., Núñez, P., Cintas, R., Calderita, L.V.: RoboComp: a tool-based robotics framework. In: Ando, N., Balakirsky, S., Hemker, T., Reggiani, M., von Stryk, O. (eds.) Simulation, Modeling, and Programming for Autonomous Robots, pp. 251–261. Springer, Berlin (2010)

Challenges on the Application
of Automated Planning
for Comprehensive Geriatric Assessment
Using an Autonomous Social Robot

Angel García-Olaya(✉)⬤, Raquel Fuentetaja⬤, Javier García-Polo,
José Carlos González⬤, and Fernando Fernández

Universidad Carlos III de Madrid, 28911 Leganés, Spain
{agolaya,rfuentet,fjgpolo,josgonza,ffernand}@inf.uc3m.es,
http://www.plg.inf.uc3m.es

Abstract. Comprehensive Geriatric Assessment is a medical procedure
to evaluate the physical, social and psychological status of elder patients.
One of its phases consists of performing different tests to the patient or
relatives. In this paper we present the challenges to apply Automated
Planning to control an autonomous robot helping the clinician to perform such tests. On the one hand the paper focuses on the modelling
decisions taken, from an initial approach where each test was encoded
using slightly different domains, to the final unified domain allowing any
test to be represented. On the other hand, the paper deals with practical
issues arisen when executing the plans. Preliminary tests performed with
real users show that the proposed approach is able to seamlessly handle
the patient-robot interaction in real time, recovering from unexpected
events and adapting to the users' preferred input method, while being
able to gather all the information needed by the clinician.

Keywords: Automated Planning · Human-Robot Interaction
Planning and execution · Social Robotics
Comprehensive Geriatric Assessment · Health-care robotics

1 Introduction

One of the main challenges of the health-care systems is the aging of the population. According to the World Health Organization [24] between 2000 and 2050,
world's population over 60 years will increase from 11% to 22%, reaching a total
of 2 billion people, from which almost 400 million will be 80 years or older.

Comprehensive Geriatric Assessment (CGA) [6] is a medical procedure to
evaluate the physical, social, cognitive and psychological status of elder patients.
CGA is the prerequisite for personalized treatment and follow-up. It is performed
usually every 6 months and involves three phases: in phase 1, Clinical Interview,
patients and relatives inform to health-care professionals about the patient status

© Springer Nature Switzerland AG 2019
R. Fuentetaja Pizán et al. (Eds.): WAF 2018, AISC 855, pp. 179–194, 2019.
https://doi.org/10.1007/978-3-319-99885-5_13

and their perception about her evolution since last evaluation. During phase 2, Multidimensional Assessments, different measurement tests, performed either by patients or relatives, evaluate the functional, cognitive, motor and social status of the patient. Finally, in phase 3, Individualized Care Plan, the results from the two previous steps are taken into account and the health-care professional designs a personalized care plan to be followed until next visit.

This paper shows the challenges faced when applying Automated Planning [11] to control a robot able to assist health-care professionals collecting information during phase 2. Currently it is able to autonomously conduct tests without health-care professional intervention. Meanwhile, the clinician can concentrate on more added-value tasks like discussing with patient or relatives or evaluating the evolution since the last visit. The use of an automatic solution to perform the tests could significantly reduce the total time spent in the CGA process.

The paper is structured as follows: Next section describes the background of the work. Section 3 shows the challenges faced and the decisions taken when developing the deliberative module. Section 4 presents a brief description of the domain and problems. In Sect. 5, preliminary results are shown. Section 6 shows the closest related work. Finally the paper conclusions and the future work are described in Sect. 7.

2   Background

2.1   CGA Tests

Several tests have been proposed in the medical literature for CGA's phase 2. Our system is currently able to perform three of the most popular ones: a functional test, the Barthel's Index Rating Scale [17]; a cognitive test, the Mini-Mental State Examination (MMSE) [8]; and a physical test, the Get Up & Go test [18]. In addition to measure different aspects of the patient status, these three tests are quite different in nature and pose different challenges from the Human-Robot Interaction (HRI) point of view:

- The Barthel's test is performed by the patient or a relative and can refer to present or past patient's conditions. It consists of ten questions, each one with three or four possible closed answers. Questions deal with the patient's ability to perform daily living activities. From the HRI perspective this is the simplest test to automatize as answers are closed and the patient just needs to answer verbally or using a tablet.
- The MMSE test evaluates cognitive impairment and changes in patients suffering from dementia. It must be performed by the patient. In 5 to 15 min it examines orientation, immediate and short-term memory, attention, calculation abilities, recall, language understanding, and ability to follow simple commands. It is more complex to automatize than the previous one as it does not only include closed-answer questions, but also open-answer questions ("What day is today?") and also questions that require monitoring simple patient movements ("Close your eyes"), painting or hand-writing.

– The Get Up & Go test is used to measure balance and fall risk, detecting deviations from a confident, normal performance. The patient must stand up from a chair, walk a short distance, turn around, return to the initial location and sit down again. The robot has not only to speak to guide the patient, but it needs also to place itself in a position where it can monitor patient movements. In addition, the robot needs to perceive the gait and analyze balance and time.

2.2 The Robot

The used robot is shown in Fig. 1. It is a mobile robot, based on the MetraLabs SCITOS G3 platform[1], with capabilities for localization, navigation and obstacle avoidance. Interaction with patients is done via speech and speech recognition, but also the embedded tablet is used to show the tests' questions and can be used by the patient to answer. For patient's convenience she is provided with a second tablet mirroring the embedded one. The robot is also endowed with a Kinect 3D sensor that detects the patient and tracks her movements, which is needed to check that she has not left the room, but also for some parts of the tests. The software and hardware description of the robot has been performed elsewhere [3] and is not part of this paper.

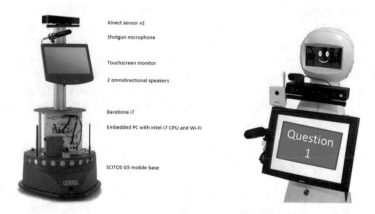

Fig. 1. The robot used for CGA, showing sensors (left) and shell (right).

2.3 Automated Planning

Automated Planning (AP) consists of finding a plan (sequence of transitions) that, when executed, takes the system from its current state (initial state) to a final one where the goals are achieved. Most of the research in AP is devoted to the automatic generation of plans using generic problem-solving techniques.

We use AP techniques to provide the robot with the ability of generating its own plans and behave autonomously when conducting the CGA tests. There are

---

[1] http://www.metralabs.com/en/ (Last visit on July 21st 2018).

two main activities that need to be done so that the application of AP is possible: First, the Human Robot Interaction (HRI) needs to be modeled; second, once the model has been created and a plan achieving the goals is found, the plan must be executed and monitored, given that the real interaction with the patient is very likely to be quite different from the "ideal" planned one.

Roughly speaking, there are two main approaches to Automated Planning: *action-based* and *timeline-based*. The main difference between them is the way transitions are modeled: *Action-based* planning [7] represents transitions by means of actions. Following this paradigm, the task is modeled in terms of states and actions; applying an action in a given state produces a new state. The plan will be the sequence of actions achieving the goal from the initial state. Formally:

Definition 1 (Action-based planning task). *An action-based planning task is a tuple* $\Pi_{ab} = \{F, A, I, G\}$.

F is a finite set of facts and numerical state variables. A state $s \subseteq F$ is defined by the set of facts that are true together with the set of values of the numerical state variables. A are the actions that the robot can perform, being each $a \in A$ described by means of four sets: $\{pre_a, add_a, del_a, num_a\}$. pre_a are the preconditions, that must hold in a state for the action to be applicable, add_a and del_a are respectively the true facts appearing and the facts that become false after a is applied, and num_a accounts for the variations that a produces in the numeric state variables. $I \subseteq F$ is the initial state and $G \subseteq F$ describes the goals. A plan, $\pi = \{a_1, a_2, ...a_n\}$ is sequence of actions that applied to I reaches a state s_n such as $G \subseteq s_n$. Classical *action-based* planning assumes instantaneous actions and does not consider time explicitly, even if it has extensions handling time.

On the other hand, *timeline-based* planning focuses on the internal state of the system, how it varies due to changes in the world and how it drives the interaction with the environment. Formally:

Definition 2 (Timeline based-planning task). *A timeline-based planning task is a tuple* $\Pi_{tb} = \{S, R, I, G\}$.

$S = \{s_a, s_b, ...s_k\}$ is set of state variables, where each $s_i \in S$ is defined by (V, T, D), being V the possible values it can take, $T : V \rightarrow 2^v$ is a transition function that for each value $v \in V$ specifies the values that s_i can take after it, and $D : V \rightarrow \mathbb{N} \times \mathbb{N}$ specifies the minimum and maximum times where s_i can take the value v. R is a synchronization rule, a pair of two state variables s_i and s_j temporarily related by one of the Allen's temporal relations [2] (*equal, before, meets, overlaps*, etc.). I is the initial value of all the state variables and G are the goals; future desired values for some of the state variables, usually temporarily tagged and with temporal relations among them. A plan is a sequence of state variables transitions that respects the synchronization rules.

In both planning paradigms modeling a task consists in creating a representation of the states and the transitions using a declarative language. *Action-based* planning uses mainly the Planning Domain Definition Language (PDDL) [19] and its variants. Meanwhile there is no standard language in *timeline-based* planning and various alternatives, like the Domain Description Language (DDL) [5] or the New Domain Definition Language (NDDL) [4], coexist. The models are usually split into two files: the domain file, which contains the definition of the predicates and numerical functions for defining states and the transitions (actions or synchronization rules); and the problem file, which contains the descriptions of the initial state and the goals. This way, a domain file can be reused to represent a family of planning tasks just by changing the problem file.

3   Automated Planning for CGA

In this section we analyze the AP challenges CGA poses from both the modelling and execution perspectives.

3.1   Modeling Challenges and Decisions Taken

There are several ways to model the CGA process using AP. Once a planning approach is selected, specific features of CGA allow to decide among the different alternatives that appear when creating the model.

Selection of the Planning Paradigm. In order to choose the paradigm the following general aspects are usually considered:

- Reasoning about time requirements: Although both paradigms are able to explicitly take time into account when creating the plans, it has been demonstrated that *timeline-based* planning is expressive enough to capture *action-based* temporal planning, while the contrary is yet unclear [13]. If complex temporal reasoning is needed, as in the case of quantitative temporal relations among goals (for example representing that a goal must be achieved between 10 and 15 s after another goal has been reached), *timeline-based* planning is usually a better approach.
- Model complexity: *Timeline-based* models tend to be more complicated than their equivalent *action-based* ones. In addition, changes in *action-based* models tend to be quite straightforward and can be performed even by non-experts, while in the another approach changes in the model usually needed to be carefully studied.
- Availability of planners: *Action-based* planning community is quite bigger than the *timeline-based* one. That means the number of freely available planners is higher and contributions in the field appear at a higher pace. Nevertheless, some applications like autonomy in space are mainly dominated by *timeline-based* approaches.

Although the concept of time is relevant in CGA interaction (for example patients have a limited time to answer before the robot repeats the question or marks it as unanswered), reasoning about time is not required, nor there is any need for concurrent actions or complex temporal relations between goals. Implicitly considering time, as non-temporal *action-based* planning does, seems a good choice, reducing the complexity of the search and producing better quality plans faster. Non-temporal models are also easier to create and there is a large amount of planners supporting it, consequently, non-temporal *action-based* planning has been selected to model the CGA interaction. As a natural result of this choice, the PDDL language will be used.

Planning Horizon. Next decision is to determine the planning horizon. We can plan in a question by question basis: the robot creates a plan to pose the current question, executes it, and once finished, looks for a new plan for the next one. Most CGA tests are episodic at the level of questions. This means each question can be planned in isolation and the result of one question has no effect in the next ones. In the rare case where some consecutive questions are interrelated they could be considered as a bigger single question. The main advantage of this approach is that it is very likely that plans will be found quickly, as they will comprise a relatively small number of actions. The drawback is that a external control system is needed to control the global interaction flow.

A second alternative is to plan the whole test, modeling each of the episodes by introducing intermediate subgoals or landmarks [16] and a mechanism to impose a total order among them. That way, an external control system is not needed, but it will take longer to find a plan. We could also think about planning the whole CGA, creating a plan for all tests to be carried during the session.

After evaluating the three alternatives, we have decided to plan at the level of a single test. As we will see in Sect. 5 planing times are short enough to guarantee a seamless interaction even when multiple replanning processes are needed. In the usual case that several tests are prescribed sequentially, a control system is needed to concatenate the planning and execution of all of them, but this is way simpler than the one needed if planning at the question level.

Model Partition. Once decided we will plan at the test level. We need to state whether the CGA interaction will be modeled using a single domain or if a domain will be created for each test.

Having a domain for each test allows to tailor them to the specific types of interactions they need; answering questions in some of them, physical activities in other ones, etc. Following this approach domains will be simpler as only the required types of interactions must be encoded. The drawback is having several similar domains, which makes changes and improvements harder. On the other hand, most tests have a common structure composed by salutation, introduction, questions and farewell. Also the order of the questions is usually fixed and most of them follow a common structure: The robot begins the interaction, usually with some kind of introduction that precedes the question. Then the question itself is posed and options, if any, are enumerated. Once this is done, it is the turn of the patient to answer or to perform some action. Questions finish with the robot giving some kind of feedback about the user action ("Thanks, we will proceed to the next question") or asking the user if the answer is right ("Is this your final answer?"). Attending to this common structure it would make sense to create a single domain to model all tests.

Considering pros and cons of each alternative, we have decided to create a single domain. This domain is able to model the different types of questions that appear in CGA. Using this unified domain, a planning problem will encode each test. Although finding a plan in this common domain is harder than if several simpler domains were considered, exploiting the sequential nature of CGA makes planning times suitable for real-time interaction.

Action Modeling: Handling Uncertainty in Actions. An *action-based* planning model comprises a description of the states and of the actions the system can perform. Actions modify the state either by changing the truth value of facts or by altering the value of numerical ones. When modeling a task using planning it is necessary to make some assumptions or predictions about the effects of the actions [12]. As we are planning the full test we need to guess which will be the patient's behavior for each question. Probabilistic planning, which extends the model of *action-based* planning assigning probabilities to each of the outcomes of an action, seems a natural approach to handle uncertainty in action execution, as for example the patient asking for the robot to repeat a question. But in practice probabilistic planning has some drawbacks: probabilities for each outcome are usually unknown and finding a plan taking them into account becomes much harder. In fact, in domains where there are no dead-ends (states from which we cannot escape, or that once reached prevent us from achieving the goals), a common procedure to handle uncertainty is to assume that the action execution will result in its most likely effect, and replan in any other case [25].

That is the approach followed in our work: A *nominal behavior*, referring to the desired flow of interaction, has been defined. It corresponds to a seamless interaction between the robot and the patient, where the first one poses the questions and the second one answers them with no errors or external events modifying the expected flow. If the interaction diverges from the predicted flow a replanning episode is launched and a new plan is created.

Action Modeling: Handling Unpredictable Events. Many things can go wrong while performing the CGA, but the ultimate goal of the robot is to find a new plan to come back to the nominal behavior in order to finish the test gathering the required information. The new plan must contain some actions to correct the unexpected issue and to return to the normal flow of the nominal behavior. These are *corrective actions*, which are never included in the initial plan. For instance, if the patient leaves the room, the robot must interrupt the test execution. The new plan should start calling the patient and searching for her before continuing with the test. After the execution of these corrective actions, the nominal behavior can continue. Modeling these corrective actions is important as it endows the system with much more responsiveness to the environment and a more coherent interaction. Dividing actions into nominal-flow ones and corrective ones simplifies both types of actions, which results in reduced planning times. Nominal behavior plus corrective actions are a good alternative to probabilistic planning approaches. They allow to easily recover from undesirable situations, scale better and do not need a probabilistic model of each action outcome to be created.

One peculiarity of CGA interaction is that often after an interruption the interaction should not come back to the point where it was when stopped, but to a previous point. For example if the robot is interrupted while reading the options of a question, it is required to start the question again. If the unexpected event occurs when waiting for the answer, it makes sense that a summary of the question is performed again when the nominal flow is resumed.

Modeling these *restart from here* states and the way to come back to them in case of interruption is somewhat special because the corrective plan must reset the values of some fluents to repeat one or more previously executed actions to achieve a coherent interaction. To solve it, from a conceptual point of view the execution has been divided in interaction episodes of different number of actions. An interaction episode must be completely executed or it has to be repeated again from its beginning. As a consequence, the number of actions to execute again after a replanning depends on the interruption but also on the moment in which it occurs.

State Modeling: Handling Numerical Information. While performing a CGA test the state must contain some numerical information, like the number of the question being currently asked, the number of times it has been repeated or how many questions the patient answered incorrectly, among others. This is represented by means of numerical functions or numerical fluents. Despite being part of the language specification since a long time ago and their obvious utility in many real applications, modern planners still have problems to integrate them in their heuristics, and indeed many state-of-the-art planners do not fully support them. A common *trick*, widely used when the range of numeric values is limited, is to codify them by using logical predicates. That way, the heuristic can consider them while looking for the solution. But this increases the number of possible states, which in turn increases the search space and the pre-processing time. In our case, where plans must be found in real time modeling numbers or order relations with predicates could increase pre-processing time to unbearable values, not suitable for a fluid interaction.

State Modeling: Types of State Variables. Two states will be different if they contain different facts or different values for the numerical state variables. Both will change as an effect of an applied action or due to external events. Taking that into account, we can conceptually divide the state variables (both logical and numerical) into three different categories:

- *Internal variables*: They are used to control the execution of the test, organize the flow of actions and specify properties of the questions of a test. Their value at any moment of the plan execution can be fully determined as they will only change after an action has been applied.

- *Predicted variables*: Although in the general case the patient can behave non-deterministically and unpredictably, we can have some expectations about her behavior while performing the test, what we called *nominal behavior*. Predicted variables represent the expected behavior of the patient during the interaction, their true value does not depend on the system, but on the patient, but planning is done assuming certain outcomes as a result of the patient's actions.
- *Unpredictable variables*: They are used to represent the effects of actions of the user that do not directly respond to an ideal interaction flow. They are not considered in the planning process as their occurrence cannot be predicted in advance. They appear as consequence of unexpected events that break the nominal flow and trigger corrective actions, whose aim is indeed to make them disappear.

This differentiation allows us to conceptually divide a state of the world into two parts: one made of variables whose values will be (almost) always as expected (internal ones), and another one comprising the variables that can change unexpectedly, invalidating the current plan in the middle of its execution (predicted and unpredictable ones).

3.2   Execution

Once the plan has been created it must be executed in the real and dynamic environment. In fact, the mechanism to join planning and execution is usually up to each developer and depends usually on the chosen architecture. In our case we are using the PELEA architecture [1]. This section describes the generic items that must be considered in executing the generated plans of actions for CGA.

Continuous Monitoring. While planning we assume actions are instantaneous since we do not need to reason about time and there are not concurrent or parallel actions. But of course actions do have duration while executing. This implies that for some preconditions it is not enough to monitor whether they hold when the action starts. It must be ensured they are kept during the whole execution. Then, we have adopted a solution similar to that of temporal *action-based* planning [9], specifying for each precondition whether it must be true at the beginning, end or during all the action execution. To simplify the search process this information is not taken into account when planning, it is only used while monitoring: the system is continuously monitoring the real state and comparing it to the expected one. This ensures the robot can react properly when something unexpected occurs.

Triggering Replanning. Different options exist in order to decide when to replan. Always replanning after a single change in the state can lead to unnecessary replanning as this change may not affect the plan, but allows also to take advantage of opportunities. Replanning only if the current plan is not valid requires a more complex monitoring mechanism, but reduces the need for replan. Finally is it also possible to replan only if the change invalidates the next action; this reduces even more the number of replanning episodes, but can lead to dead ends. In our domain there are no opportunities as the nominal behavior plan needs always to be followed, being the shorter path to reach the goals. Nor there are dead ends as it is always possible to find a plan that finishes the test, either successfully or not. Taking that into account we opted for the last approach, replanning only if the next action cannot be applied or if the preconditions of the current action are invalidated during its execution.

Interrupting Actions. A true interactive system must be responsive in any moment of its execution. This is especially important while it is executing an action but it has not finished yet. For example, if the robot is executing a Say action, which involves a 20 s speech and the patient leaves the room in the fifth second, the robot must be able to interrupt the speech in the middle of its execution. If not, it would continue speaking to nobody during the next 15 s before realizing that it has to call the patient to return to the test area.

Planning Time Restrictions. After an unexpected event, the robot stops in the middle of the current interaction while replanning, as it does not know the next action until the new plan is created. This imposes strict planning time restrictions: planning must be performed so there is no detectable delay in the interaction. It is very important to keep replanning times as low as possible to achieve a fluid interaction. Replanning times higher than 3 s are considered non-assumable for a proper and seamless interaction.

4   Example of Domain and Problems

In a first version of the work, a different domain was created for each of the tests. Actually, a first domain was designed to conduct the Barthel's test, which was later updated to be able to cope with MMSE and Get Up & Go tests. As a result three similar but slightly different domains were created, which resulted in a kind of *spaghetti code*. Although preliminary tests with real patients have been conducted using those domains, updates and maintenance became a problem. For that reason, taking into account the learned lessons, we designed a second unique domain, able to represent the three tests. Each test will be instantiated using a different problem.

The domain contains two types of actions, those that conform the nominal flow and the corrective ones that return to it in case of unexpected events.

Figure 2 shows the PDDL description of the *Say* action, the parameters are the current question and the specific label inside the question that identifies what will be said. In the preconditions, (1) accounts for all the external events, if any appears, this fact will be false preventing the action from being executed.

```
(:action say
   :parameters   (?q - question ?l - label)
   :precondition   (and (can_continue)                          (1)
      (test_configured)                                         (2)
      (= (question_position ?q) (current_question))            (3)
      (belongs ?l ?q)                                          (4)
      (= (label_order ?l) (current_label))                    (5)
      (< (repetitions ?l) (max_repetitions ?l))               (6)
      (not (answer_received ?q))                              (7)
      (not (waiting_for_behavior ?l))                         (8)
      (not (answer_needed))                                   (9)
      (not (pending_confirmation)))                          (10)
   :effect   (and   (increase (repetitions ?l) 1)
      (when (needs_feedback ?l) (answer_needed))
      (when (not (needs_feedback ?l)) (increase (current_label) 1)))))
```

Fig. 2. Example of an action: the *Say* action.

Next preconditions are used to check the current question and the current label to be said (3–5), while controlling the maximum number of repetitions for this question has not been achieved yet, nor the answer has been already received (6–7). Finally we check too we are not waiting for the robot to perform any action (8), or for the patient to answer (9) or to confirm a previously entered answer (10). In the effects, the number of repetitions is increased. If after this action

(= (question_order q1) 3) ← the position of the question in the test
(behavior_of_event q1 question_failed ignore) ← what to do if an answer is not provided (ignore means the question is marked for later answer)
(behavior_of_event q1 doctor_needed call_doctor) ← what to do if doctor needed
(behavior_of_event q1 patient_absent call_patient) ← what to do if the patient is not visible
(behavior_of_event q1 max_q_failed call_doctor) ← what to do if after failing this question the max number of failed questions is reached
(behavior_of_event q1 max_a_failed ignore) ← what to do if the patient fails max_failed_answers in this question (notice that the system can try max_repetitions to get an answer before marking it as missing)
(= (answers q1) 0) ← the number of answers received for this question
(= (answers_required q1) 1) ← the number of answers this question requires
(= (number_failed_answers q1) 0) ← the number of non-provided answers
(= (max_failed_answers q1) 1) ← the maximum number of non-provided answers for this question before the max_a_failed event is triggered
(belongs q1_s1 q1) ← the label
(= (label_order q1_s1) 0) ← this is the first label
(= (max_repetitions q1_s1) 1) ← this label will be repeated at most once
(= (repetitions q1_s1) 0) ← the number of times this label has been repeated

Fig. 3. A partial example of the definition of a question of the Barthel's test.

the patient has to answer, a fact is inserted to account for it, if not, we continue with the question. Figure 3 shows how a question is encoded in a problem.

5   Preliminary Results

A preliminary evaluation was conducted in a retirement home in Seville, Spain, on November 2017 involving 8 patients. It was shown the robot is able to seamlessly conduct the CGA for the three selected tests, although many suggestions were made by the clinicians. Among many other improvements and changes, this resulted in a modification of the planning domains and problems. Practical reasons forced us to do those modifications while creating in parallel the unified domain. Prior to it inclusion in the real robot, severe testing in simulation has been performed to verify both domains are equivalent and the unified domain can be used in real practice with no changes in the rest of the system.

In this section we compare in simulation the planning and replanning times of both the original domain and the unified domain for the Barthel's and Get Up & Go tests. In the simulations, there is a 10% chance that the robot loses track of the patient. Additionally, in Barthel test, we have included an interaction error for each question (i.e., the robot does not receive the answer from the patient, or the patient does not respond) with a probability of 30%. Finally, in Get up & Go test, we have included different detection errors: the robot does not detect the patient near the chair (10%), it does not detect the patient seated (20%), and it detects the patient has walked more than the required distance (10%). Experiments were conducted on a 64 bits Intel Xeon 2.93 GHZ Quad Core processor with 2 GB RAM, using Linux and the Metric-FF planner [15].

Table 1 provides the results for each domain and test, analyzed across five dimensions: the accumulated time (seconds) used to perform the test , the number of actions in the executed plan, the number of times it is necessary to replan, the average time (seconds) needed to build a plan, and the average time needed to start executing each action (milliseconds). Means and standard deviations computed after ten different executions for each domain and test are shown.

Table 1. Results in Barthel and Get up & Go tests for the unified and specific domains.

Test	Domain	Time (s)	# Actions	# Replans	Planning (s.)	Response (ms)
Barthel	Unified	723.2 ± 42.9	123.5 ± 8.4	6.3 ± 2.8	0.08 ± 0.01	281.5 ± 5.3
	Specific	716.5 ± 64.5	102.0 ± 5.3	5.8 ± 3.3	0.33 ± 0.01	277.7 ± 6.8
GetUpGo	Unified	167.9 ± 8.9	33.0 ± 2.7	3.0 ± 0.9	0.01 ± 0.00	232.5 ± 4.9
	Specific	166.7 ± 9.2	20.0 ± 1.8	2.2 ± 0.7	0.05 ± 0.03	236.1 ± 5.6

The unified domain requires a higher number of actions to solve the tests, however, the accumulated time needed to perform each test is just slightly affected by this increment in the number of actions. Most of these additional actions control the internal flow of the planning process and they do not have associated low-level actions to be performed on the robot. The planner finds plans much faster using the unified domain than the specific one. The unified domain simplifies the previous ones, containing just the information needed to reason and to drive the interaction between the patient and the robot (e.g., pauses between speech segments are not planned). The number of replanning episodes is not affected by the use of one or other domain and less than 250 ms. are needed to start executing each action.

6   Related Work

The use of Automated Planning for language and dialog generation, which is an important part of social interaction, goes back to the early 80s [20], but these efforts suffered from the poor performance of planners available that time. More recently, AP has been used for situated natural language generation, by encoding a PDDL problem that can be solved by any planner [10]. Instead of generating the language, our phrases are prerecorded and is the interaction flow what is generated using planning. In [22] domain-independent planning is used to generate conversations. They use a planner able to handle incomplete information and sensing actions by means of a non-standard representation language, which can be compiled to PDDL. The planner divides the information into five different databases and reasons to generate new knowledge that is stored into them. A similar approach, using the same planner, is used to perform action selection in a robot bartender scenario [21]. In particular, the task of interacting with human customers is mixed with the physical task of ensuring that the correct drinks are delivered to the correct customers. Conditional plans are generated if the needed information is not available yet. In the case of unexpected situations a new plan is generated. This application is quite similar to ours as both interaction and robot behavior must be planned, but we use standard PDDL, with no conditional plans, being able to use any planner released by the community.

The STRANDS project [14] aims to provide long-term autonomy to indoor robots, using also the MetraLabs SCITOS platform. One of the scenarios where their solutions has been deployed has been a large elder-care facility, where it acted as an information point, guided visitors and helped in walking-based therapies. Unlike in our case tasks are considered to be atomic so they are scheduled, not planned. An interesting feature, which we would like to explore in the future, is the robot ability to learn from experience. It is able to predict future states (for example if a door will we open at certain times), update traverse times and success ratio (using a Markov Decision Process), or learn user interaction patterns (probability of users wanting to interact at given times and locations).

Action-base Planning, Timeline-based Planning and Constrain-based Scheduling have been compared to control multiple robots deployed to assist elderly residents in a retirement home [23]. For that specific scenario, Constraint-based scheduling seems to be the most appropriate technique: PDDL-based planning finds always low quality plans, while Timeline-based planning is unable to model the problem unless the solver is modified accordingly. Their approach involves multiple robots in a temporal problem that lies in the intersection of planning and scheduling. Our problem only involves a single robot, thus temporal aspects and concurrency are not imperative, and given the sequential nature of tests, plans are always optimal or near optimal. Scheduling of tests is currently done by the healthcare professionals, and although we plan to provide automatic mixed-initiative scheduling capacities to our system, scheduling a test and running it are to some extent independent and different techniques can be used for each of them.

7   Conclusions and Future Work

In this paper we have presented the decisions taken while modeling CGA interaction and showed that using the created domain and problems we can find plans able to control the robot while performing the Barthel's and Get Up & Go tests. Plans are found in less than a tenth of second, which fully meets the real time requirements for CGA interaction. The domain has been created to follow a nominal behavior of the patient. In the likely case that the patient does not stick to it, we use replanning to come back to the expected flow, and again this takes less than 0.1 s.

In a future we want to use the unified domain to encode also the MMSE test and any other test required by the clinical experimentation. A large scale pilot with real patients will be carried out in fall 2018 in retirement homes at Seville, Spain and Troyes, France.

Acknowledgment. This work has been partially funded by the European Union ECHORD++ project (FP7-ICT-601116) and the TIN2015-65686-C5 Spanish Ministerio de Economía y Competitividad project. Javier García is partially supported by the Comunidad de Madrid (Spain) funds under the project 2016-T2/TIC-1712.

References

1. Alcázar, V., Guzmán, C., Prior, D., Borrajo, D., Castillo, L., Onaindia, E.: PELEA: planning, learning and execution architecture. In: Proceedings of the 28th Workshop of the UK Planning and Scheduling Special Interest Group (PlanSIG), Brescia, Italy (2010)
2. Allen, J.: Maintaining knowledge about temporal intervals. Commun. ACM **26**(11), 832–843 (1983)
3. Bandera, A., et al.: CLARC: a robotic architecture for comprehensive geriatric assessment. In: Workshop on Physical Agents, pp. 1–8 (2016). ISBN 978-84-608-8176-6
4. Barreiro, J., Boyce, M., Do, M., Frank, J., Iatauro, M., Kichkaylo, T., Morris, P., Ong, J., Remolina, E., Smith, T., et al.: EUROPA: a platform for AI planning, scheduling, constraint programming, and optimization. Technical report, 4th International Competition on Knowledge Engineering for Planning and Scheduling (ICKEPS) (2012)

5. Cesta, A., Oddi, A.: DDL.1: a formal description of a constraint representation language for physical domains. In: Ghallab, M., Milani, A. (eds.) New Directions in AI Planning. IOS Press, Amsterdam (1996)
6. Ellis, G., Langhorne, P.: Comprehensive geriatric assessment for older hospital patients. Br. Med. Bull. **71**(1), 45–59 (2005). https://doi.org/10.1093/bmb/ldh033
7. Fikes, R., Nilsson, N.J.: STRIPS: a new approach to the application of theorem proving to problem solving. Artif. Intell. **2**(3/4), 189–208 (1971)
8. Folstein, M., Folstein, S., McHugh, P.: Mini-mental state: a practical method for grading the cognitive state of patients for the clinician. J. Psychiatr. Res. **12**, 189–198 (1975)
9. Fox, M., Long, D.: PDDL2.1: an extension to PDDL for expressing temporal planning domains. J. Artif. Intell. Res. (JAIR) **20**(1), 61–124 (2003)
10. Garoufi, K., Koller, A.: Automated planning for situated natural language generation. In: Proceedings of the 48th Annual Meeting of the Association for Computational Linguistics, pp. 1573–1582 (2010)
11. Ghallab, M., Nau, D., Traverso, P.: Automated Planning: Theory and Practice. Elsevier, Amsterdam (2004)
12. Ghallab, M., Nau, D., Traverso, P.: The actor's view of automated planning and acting: a position paper. Artif. Intell. **208**, 1–17 (2014)
13. Gigante, N., Montanari, A., Mayer, M.C., Orlandini, A.: Timelines are expressive enough to capture action-based temporal planning. In: 23rd International Symposium on Temporal Representation and Reasoning (TIME), pp. 100–109. IEEE (2016). https://doi.org/10.1109/TIME.2016.18
14. Hawes, N., Burbridge, C., et al.: The STRANDS project: long-term autonomy in everyday environments. IEEE Robot. Autom. Mag. **24**(3), 146–156 (2017). https://doi.org/10.1109/MRA.2016.2636359
15. Hoffmann, J.: The metric-FF planning system: translating "ignoring delete lists" to numeric state variables. J. Artif. Intell. Res. (JAIR) **20**(1), 291–341 (2003)
16. Hoffmann, J., Porteous, J., Sebastia, L.: Ordered landmarks in planning. J. Artif. Intell. Res. (JAIR) **22**, 215–278 (2004)
17. Mahoney, F., Barthel, D.: Functional evaluation: the barthel index. Maryland State Med. J. **14**, 56–61 (1965)
18. Mathias, S., Nayak, U., Isaacs, B.: Balance in elderly patients: the get-up and go test. Arch. Phys. Med. Rehabil. **67**, 387–389 (1986)
19. McDermott, D.: PDDL - the planning domain definition language. Technical report. CVC TR-98-003/DCS TR-1165, Yale Center for Computational Vision and Control (1998)
20. Perrault, C.R., Allen, J.F.: A plan-based analysis of indirect speech acts. Comput. Linguist. **6**(3–4), 167–182 (1980)
21. Petrick, R.P.A., Foster, M.E.: Planning for social interaction in a robot bartender domain. In: Proceedings of the Twenty-Third International Conference on Automated Planning and Scheduling, pp. 389–397 (2013)
22. Steedman, M., Petrick, R.P.A.: Planning dialog actions. In: Proceedings of the 8th Workshop on Discourse and Dialogue (SIGDIAL), pp. 265–272 (2007)
23. Tran, T.T., Vaquero, T., Nejat, G., Beck, J.C.: Robots in retirement homes: applying off-the-shelf planning and scheduling to a team of assistive robots. J. Artif. Intell. Res. **58**, 523–590 (2017)

24. World Health Organization: Ageing and life-course. Facts about ageing (2014). http://www.who.int/ageing/about/facts/en/
25. Yoon, S.W., Fern, A., Givan, R.: FF-Replan: a baseline for probabilistic planning. In: International Conference on Automated Planning and Schedulling (ICAPS), pp. 352–359 (2007)

Planning Human-Robot Interaction for Social Navigation in Crowded Environments

Araceli Vega[1], Luis J. Manso[2], Ramón Cintas[1], and Pedro Núñez[1(✉)]

[1] RoboLab, Escuela Politécnica, Universidad de Extremadura, Cáceres, Spain
pnuntru@unex.es
[2] School of Engineering and Applied Science, Aston University, Birmingham, UK
http://robolab.unex.es

Abstract. Navigation is one of the crucial skills autonomous robots need to perform daily tasks, and many of the rest depend on it. In this paper, we argue that this dependence goes both ways in advanced social autonomous robots. Manipulation, perception, and most importantly human-robot interaction are some of the skills in which navigation might rely on. This paper is focused on the dependence on human-robot interaction and uses two particular scenarios of growing complexity as an example: asking for collaboration to enter a room and asking for permission to navigate between two people which are talking. In the first scenario, the person physically blocks the path to the adjacent room, so it would be impossible for the robot to navigate to such room. Even though in the second scenario the people talking do not block the path to the other room, from a social point of view, interrupting an ongoing conversation without noticing is undesirable. In this paper we propose a navigation planning domain and a set of software agents which allow the robot to navigate in crowded environments in a socially acceptable way, asking for cooperation or permission when necessary. The paper provides quantitative experimental results including social navigation metrics and the results of a Likert-scale satisfaction questionnaire.

1 Introduction

A future where humans and robots coexist appears to be getting increasingly close. In fact, some applications in which social robots help humans in their daily tasks already exist. For instance, social robots in therapy or education have proven feasible and successful in use [1]. Other social robots are being developed to provide the elderly with assistance at home or in nursing homes, and even to provide health specialists with help during their working hours. Many of these applications for robots require them to work alongside people as capable and socially smart partners.

In these scenarios, navigation is one of the most important tasks social robots need to perform. In fact, mapping, localisation and path planning, which are the foundations of robot navigation, have been among the most significant research

© Springer Nature Switzerland AG 2019
R. Fuentetaja Pizán et al. (Eds.): WAF 2018, AISC 855, pp. 195–208, 2019.
https://doi.org/10.1007/978-3-319-99885-5_14

lines for years. The field of social navigation is experiencing a remarkable growth because in environments with humans where some of the elements (objects and people) are dynamic, humans' comfortability, safety and intentions must be prioritised. Social navigation adapts robot navigation to scenarios with people by following social norms. For instance, robots should avoid getting to close to people or disturbing people who are not willing to interact with them.

In our previous work [2], a social path planner for modelling robot navigation in populated environments was proposed. In [3,4] an algorithm for human-centred navigation where a novel method for clustering groups of people in the robot's surrounding was used. According to these clusters, a social map was defined. Other works such as [5] also generate similar maps. Unfortunately, sometimes avoiding disturbing people while navigating makes impossible for the robot to reach its desired destination (*e.g.*, when a person blocks a door, or when the only path would require the robot to navigate between people which are interacting). Current algorithms have serious problems to find solutions, and frequently social robots cannot find a way to reach their destinations. The paper at hand extends previous works [3,4] and focuses on a social planning strategy for situations where people block the path and there are no alternative paths.

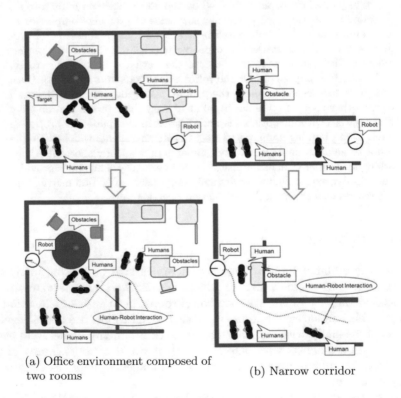

(a) Office environment composed of two rooms

(b) Narrow corridor

Fig. 1. Examples of scenarios in which robots have navigate using social rules.

Consider the cases illustrated in Fig. 1. In both images the robot must navigate in human-populated scenarios from its initial position to a target position, and there are blocked areas in the routes planned. In Fig. 1a the human next to the door blocks the path; in a similar scenario, in Fig. 1b two people are interacting and the social map generated by our clustering algorithm also blocks the robot path. As the **main contribution** of this paper, we propose a social navigation planning domain and a set of software agents which allow robots to navigate in human-populated environments in a socially acceptable way, asking for permission or cooperation when necessary.

This paper is organized as follows: after discussing known approaches to human-centered navigation for robot navigation in Sect. 2 and Sect. 3 presents the cognitive architecture CORTEX, which consists of a network of software agents that allows executing complex tasks involving skills such as human perception or robot navigation. In Sect. 4 we describe the proposed navigation planning domain. Section 5 presents the experimental results. Section 6 describes the conclusions drawn and future work.

2   Background

The debate between the use of grid-based vs. topological maps [6] has existed, including on whether or not using maps at all. While robots have been able to perform rather complex tasks without representations of any kind, when robots are supposed to perform them efficiently, map-less approaches becomes hard to support. Considerably rich and structured world models, with a higher *semantic* load than two-dimensional grids became frequently necessary to use in these cases. Suggesting scenarios in which a structured representation is necessary is not hard, just consider a dialogue between a robot and a person; interpreting human commands such as "pick up the red ball and bring it to my sister" is a good example. The robot would require information about kinship and the balls that they have seen (there might be more than one). This kind of information is required for almost all HRI skill. *Social mapping*, was introduced in [5]. It deals with the problem of human-aware robot navigation and considers factors like human comfort, sociability, predictability, safety and naturalness [7]. More recently, the concept of *behavioral mapping* has been introduced in [8], where the authors extend social mapping to a behavioral model acting as a mediator that facilitates seamless cooperation among the humans.

Robot navigation in crowded environments has been extensively studied in the last years and several theories and methods have been proposed since then. Particularly interesting reviews have been presented in [7,9] and more recently [10]. Classic social navigation paradigms are based on using well-known navigation algorithms, and therefore adding social conventions and/or social constraints. Under this prism, different works such as [11,12], have shown that the same proxemic zones that exist in human-human interaction can also be applied to human-robot interaction scenarios. A broad survey and discussion regarding the social concepts of proxemics theory applied in the context of human-aware autonomous navigation was presented in [9]. A proxemic-based adaptive spatial density function was defined for clustering groups of people and define forbidden spaces in [3].

As the number of skills robots have increases human-robot collaboration becomes more feasible. In [13], the requirements for effective human-robot collaboration in interactive navigation scenarios are listed. Additionally, authors present three different human-robot collaborative planners. However, they only focus on secure navigation and not on HRI. Other works such as [14], anticipate the human trajectory in order to update social constraints during robot navigation. Similar works are presented in [15]. Again, authors do not take into account interaction with people for robot navigation. Planning for HRI has been used in manipulation tasks [16] and task allocation in collaborative industrial assembly processes. However, there are no works where HRI has been used to improve robot navigation in crowed environment using social conventions. This paper introduces a planning domain for social navigation where HRI is crucial for solving real situations where the robot's path is blocked due to social limitations. The goal is for the robot to execute actions that optimise social navigation and human satisfaction.

3   Cognitive Architecture for Social Navigation

To properly understand the proposal at hand it is necessary to familiarise with CORTEX, the cognitive architecture used [17]. Social robotics systems are getting more and more complex: different robotic skills are needed in order to achieve the tasks that robots are currently expected to do. The robotics cognitive architecture CORTEX is defined structurally as a network of cooperative *software agents* connected through a *shared representation* (see Fig. 2). This shared representation was defined in [17], as *"a directed multi-labelled graph where nodes represent symbolic or geometric entities and edges represent symbolic or geometric relationships"*.

In the proposal of a flexible and adaptive spatial density function for social navigation, different CORTEX agents are involved. First, in the higher layer of the architecture the robot must have the capability of detecting objects in the path and updating the symbolic model accordingly. Additionally, the skill of detecting humans is also mandatory because robots need to know about humans to get commands, avoid collisions and provide feedback. The final, and most important agent for social navigation, is the one implementing the navigation algorithms that allows robots to navigate from a point to another in a secure and social manner (implementation of the path-planning, localization and SLAM algorithms, among other).

3.1   Deep State Representation

The concept of *deep representations* was initially described by Beetz et al. [18] and it advocates the integrated representation of robot's knowledge at various

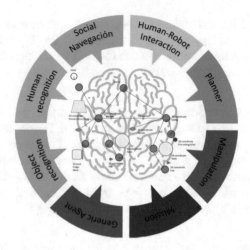

Fig. 2. Diagram of CORTEX with the main software agents involved in this work. The shared representation of the environment is represented in the centre.

levels of abstraction in a unique, articulated structure such as a graph. Based on this concept, a new shared representation, *Deep State Representation (DSR)*, to hold the robot's belief as a combination of symbolic and geometric information, is proposed in [17]. This new structure represents knowledge about the robot itself and the world around it in a flexible and scalable way.

3.2 Agents

An agent within CORTEX is defined as *a computational entity in charge of a well-defined functionality, whether it be reactive, deliberative of hybrid, that interacts with other agents inside a well-defined framework, to enact a larger system.* In CORTEX, agents define the classic functionalities or skills of cognitive robotics architectures, such as navigation, manipulation, person perception, object perception, dialogue, reasoning, planning, symbolic learning or executing. These agents operate in a goal-oriented regime and their goals can come from outside through the agent interface, and can also be part of the agent normal operation. Next, a description of the main software agents needed for social navigation is shown:

Human Detection and Representation. The person detector agent responsible for detecting and tracking the people in front of the robot. Humans do not usually want their personal space being invaded by robots. The presence of humans in the robots' path or in their environment may determine changes in the navigation route in order to make it socially acceptable. The person detector agent acquires the information using an RGBD sensor. For each detected person the agent inserts in the DSR the pose of its torso, its upper limbs, and the head.

The lower limbs are ignored because they do not provide as much social information as the head, the upper limbs and the torso do [17]. The torso is used to avoid entering the personal space of humans and as an indicator of the possible directions in which they might walk.

Human-Robot Interaction. The conversation agent performs speech-based human-robot interaction. In social environments, HRI provides tools to the robot and/or human to communicate and collaborate. Therefore, this agent is used to include information in the model when humans tell robots about unknown objects and to properly acquire commands. Automatic Speech Recognition and Text-to-Speech algorithms allow robot to send and receive information to/from humans during its social navigation.

Executive. The Executive is responsible for computing plans to achieve the current mission, managing the changes made to the DSR by the agents as a result of their interaction with the world, and monitoring the execution of the plan. The active agents collaborate executing the actions in the plan steps as long as they consider them valid (it must be taken into account that agents might have a reactive part). Each time a structural change is included in the model, the Executive uses the domain knowledge, the current model, the target and the previous plan to update the current plan accordingly. The Executive agent is able use different planners. Currently AGGL [19] and PDDL-based [20] planners are supported.

Social Navigation. Navigation is in charge of performing local navigation, complying with social rules and including the location of the robot in the DSR. Global path planning is performed by the symbolic planner used by the executive.

The social navigation algorithm implemented is an evolution of the work presented in [3]. Currently, the Dijkstra algorithm is used to determine the shortest path between the position of the robot and its target, and the elastic band algorithm [21] is used to optimize the trajectory.

The robot calculates the initial trajectory based on a free space graph formed by occupied and free points. Each point has a social cost that represents how inconvenient it is for humans to see the robot go through each point. This cost is used to weight the edges of the graph. The cost of a path is the sum of the costs of the points that compose it. The navigation algorithm uses the Dijkstra algorithm to find the shortest route between the points of the graph, based on the cost of the path. Once the path is computed, the robot optimizes the trajectory using the elastic bands algorithm, which consists in the calculation of attraction and repulsion forces to get the robot away from the obstacles on the road.

In environments with humans, the robot creates a social map based on the description of the personal space of the humans present. Three social zones have been defined around the person: intimate zone, personal zone and social zone. These spaces were introduced in [22]. The free space graph is adapted to these zones by modifying the costs of the points contained in said areas. By increasing the cost of the personal and social zones, robots will try to avoid these areas in the planning of the shortest route. The intimate zone is considered a forbidden setting as occupied the points of the graph contained in that area. Figure 3 shows a free space graph and the defined zones, characterized by asymmetric Gaussian curves.

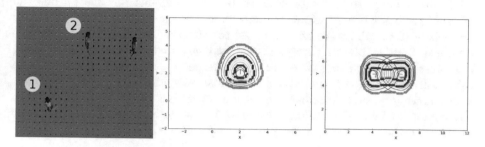

Fig. 3. Free space graph and the social zones defined. In colour red is shown the intimate area, in purple the personal area and in colour blue the social one. Number one shows the social zones for an individual human and number two shows the cluster of two persons interacting.

If there is an interaction between the humans present, the planner groups them together, in such a way that it is forbidden to pass between them, ensuring that the robot does not interrupt the interaction.

Regarding localization algorithms, the navigation agent is algorithm-independent. It has been used with different ones with different properties, so the algorithm can be changed depending on the characteristics of the environment.

4   Planning Human-Robot Interaction

As in any other context, planning human-aware navigation tasks entails defining the elements of the planning problem: an initial world model, a mission, and a set of actions (*i.e.*, the planning domain). In traditional automated planning, the initial world model is composed of a set of symbols and a set of n-ary predicates that are used to provide information regarding such symbols. In CORTEX, planning is performed similarly with the symbolic information in the DSR, using the nodes of the representation as symbols and the edges of the graph as predicates. The fact that the predicates in CORTEX are limited to the edges in the DSR limits the usage of predicates to unary and binary ones, but it also facilitates the visualization of the symbolic world model representation [19]. The remaining of this section describes the symbols and predicates (nodes and edges used in the DSR) that support estimating the best plan to make the robot achieve its navigation tasks. In this paper the term predicate and edge will be used indistinctly.

4.1   Symbols and Predicates

For navigation purposes, the robot uses three types of symbols: *human, robot,* and *room*. This paper investigates the case where only a *robot* is found in the model, but the existence of several humans and rooms is possible.

The robot and each person must be located within an existing room; for this purpose an *in* predicate (edge in the DSR) is used. Robots and humans might be paying attention to other robots and humans; for this purpose an *interact* predicate is used. Humans might block the path of the robot, physically or socially (*i.e.*, robots are not supposed to interfere visually when two people interact). Physical blocking is represented using *block* edges, while social blocking is represented using *softblock* edges. To represent that a robot is close enough to establish social interaction with a human the robot includes *reach* predicates. The following section describes the most relevant actions of the human-aware navigation domain. The actions described in the domain will help the reader understand the meaning and usage of these predicates.

4.2   Navigation Domain

The whole navigation domain is composed of 12 actions. The three most important are described in this section: engageHuman, askForPermissionToPass, and askForCollaborationToPass.

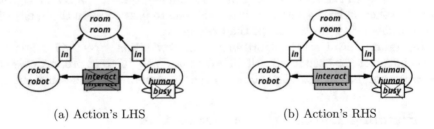

(a) Action's LHS   (b) Action's RHS

Fig. 4. Action *engage*.

engageHuman. In this proposal the first step to ask for help or permission when navigating is engaging human interaction. This very goal is the purpose of the *engageHuman* action. The action (see Fig. 4) states that if a human is reachable the robot can interact with such person unless its symbol is marked as busy (which is done when a human explicitly says that she or he does not want to be disturbed). The effect of the correct execution of the action is two new *interact* edges, one from the human to the robot and *vice versa*.

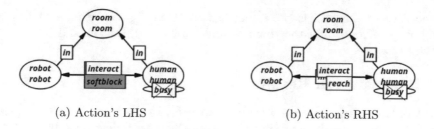

(a) Action's LHS   (b) Action's RHS

Fig. 5. Action *askForPermissionToPass*.

askForPermissionToPass. Once the robot is interacting with a particular human it can ask for permission to pass in the case it needs to cross the viewpoint of two humans which are interacting among themselves. This would also apply in other use cases when, for example, humans are watching television. The precondition for the robot to be able to execute the *askForPermissionToPass* action would be the existence of a human blocking its path *socially* (predicate *softblock*) in the same room where it is located with whom the robot should be interacting. The outcome of the successful execution of the action is that the human stops blocking the way of the robot socially. See Fig. 5 for the visual definition of the action.

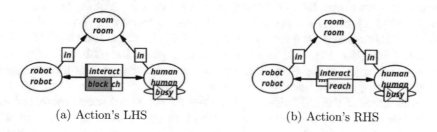

(a) Action's LHS   (b) Action's RHS

Fig. 6. Action *askForCollaboration*.

askForCollaboration. The action *askForCollaboration* is similar to the previous action *askForPermission*. The only difference is that in this case the human blocks the way physically and not socially. Therefore, in this case the robot asks the human to move and waits. See Fig. 6 for the visual definition of the action.

4.3   Missions

The missions are defined as in other systems, describing a subset of existing symbols and the predicates that should be true. The two types of experiment performed require the robot to go from one room (id 3) to an adjacent room (room id 5), even though there are people blocking its way physically or socially.

In both cases the goal is the same, "go to room 5", so the mission has two symbols, the robot and the room 5, and a predicate that should be true *(in robot_1 room_5)*.

5 Experimental Results

A set of simulated scenarios were used to validate the results of the proposed navigation planning domain. The algorithms have been developed in C++ and the tests have been performed in a PC with anIntel Core i5 processor with 4 Gb of DDR3 RAM and Ubuntu GNU/Linux 16.10. We evaluate both, quantitative and qualitative experimental results, including social navigation metrics and the results of a Likert-scale satisfaction questionnaire. We use a simulated version of the robot Viriato, a social robot equipped with an RGBD camera and laser range sensor.

5.1 Description of the Experiments

The simulated scenario is a $65\,\mathrm{m}^2$ two-room apartment equipped with a kitchen and a living room, where two different tests are described[1]: (i) First, a human blocks the path in the corridor; and (ii) two people talk in a vis-a-vis formation blocking the robot path. The robot Viriato navigates through this apartment to several positions. Figure 7 summarises the tests in six steps: In Fig. 7.1 the robot starts its route. Its first target is located in the corridor. In Fig. 7.2 the robot plans its path and navigates to the human. After asking for collaboration, the robot navigates to the first target (Fig. 7.3). In the second test (Fig. 7.4), the robot has the target in the second room, plans its path and initiates a conversation with people (Fig. 7.5). Finally, once the robot asks for permission to pass, it navigates to its target position.

An example of the HRI planning is shown in Fig. 8: our social robot (labeled as '1') has an approach behavior with which it can initiate conversation with people (labeled as '2'). As the path is blocked, the robot asks for cooperation (labeled as '3'). Once the path is free, our social robot navigates until its target (labeled as '4'). A zoom of this test in '2' and '3' robot position is drawn on the right, where the changes in the graph of free space are illustrated.

In order to assess the effectiveness of the proposed navigation approach, the methodology has been evaluated accordingly to these metrics in both static scenarios: (i) average minimum distance to a human during navigation, d_{min}; (ii) distance traveled, d_t; (iii) navigation time, τ; (iv) cumulative heading changes, CHC; and (v) personal space intrusions, Ψ. These metrics have been already established by the scientific community (see [8,23]). Results are summarised in Tables 1 and 2.

To assess the satisfaction of the humans regarding the robot's behavior and HRI abilities, a Likert scale-based questionnaire was provided to a total of 34 participants. The results of the questionnaire, including the questions are shown in Table 3.

---

[1] A video of the experiments is located on goo.gl/KdGYBN.

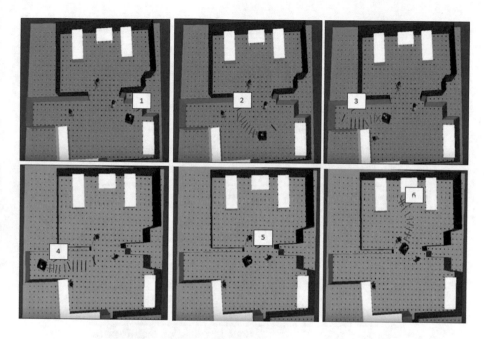

Fig. 7. The tests used in this paper in six steps

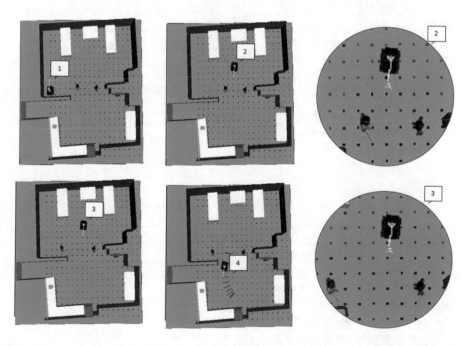

Fig. 8. An example of the HRI planning described in this paper: ask for collaboration.

Table 1. First experiment

Parameter	Social navigation architecture
d_t (m)	11.68 (2.00)
τ (s)	46.06 (14.75)
CHC	0.76 (0.15)
d_{min} Person 1 (m)	10.26 (0.54)
d_{min} Person 2 (m)	11.01 (0.42)
d_{min} Person 3 (m)	11.96 (0.39)
Ψ (Intimate) (%)	0.0 (0.0)
Ψ (Personal)(%)	0.0 (0.0)
Ψ (Social + Public)(%)	100.0 (0.0)

Table 2. Second experiment

Parameter	Social navigation architecture
d_t (m)	10.58 (1.75)
τ (s)	44.8 (12.52)
CHC	1.37 (0.18)
d_{min} Person 1 (m)	8.94 (0.58)
d_{min} Person 2 (m)	14.59 (1.14)
d_{min} Person 3 (m)	10.64 (1.31)
Ψ (Intimate) (%)	0.0 (0.0)
Ψ (Personal)(%)	0.0 (0.0)
Ψ (Social + Public) (%)	100.0 (0.0)

Table 3. Satisfaction questionnaire.

Question	avg. answer (σ)
Robot's behavior is socially appropriate in exp. 1	4.41 (0.54)
Robot's behavior is socially appropriate in exp. 2	4.47 (0.40)
Robot's behavior is friendly in exp. 1	3.79 (0.60)
Robot's behavior is friendly in exp. 2	4.05 (0.52)
The robot understands the social context and the interaction in exp. 1	4.32 (0.62)
The robot understands the social context and the interaction in exp. 2	4.37 (0.71)

6   Conclusions and Future Works

This paper provided a detailed introduction to the problem which autonomous navigation aims to solve, with a special emphasis on human-aware navigation. Despite there are many approaches to human-aware navigation, this is the first work focused on planning navigation tasks in collaboration with humans taking human social rules into account (see the *askForPermission* action).

This paper provided qualitative and quantitative results for the experiments conducted. The quantitative data support the claim that the robot does not interfere with humans, keeping a good distance and navigating properly (*e.g.*, a small Cumulative Heading Changes -CHC- value). The results of the questionnaire evince that humans are satisfied with the overall robot's behavior.

Acknowledgments. This work has been partially supported by the MICINN Project TIN2015-65686-C5-5-R, by the Extremaduran Goverment project GR15120, by the Red de Excelencia "Red de Agentes Físicos" TIN2015-71693-REDT, and by the FEDER project 0043-EUROAGE-4-E (Interreg V-A Portugal-Spain - POCTEP).

References

1. Heerink, M., Vanderborght, B., Broekens, J., et al.: Int. J. Soc. Robot. **8**, 443 (2016). https://doi.org/10.1007/s12369-016-0374-7
2. Núñez, P., Manso, L.J., Bustos, P., Drews-Jr, P., Macharet, D.G.: A proposal for the design of a semantic social path planner using CORTEX. In: Workshops on Physical Agent, pp. 31–37 (2016)
3. Vega, A., Manso, L.J., Macharet, D.G., Bustos, P., Núñez, P.: A new strategy based on an adaptive spatial density function for social robot navigation in human-populated environments. In: Proceedings of REACTS workshop at the International Conference on Computer Analysis and Patterns (2017)
4. Vega, A., Manso, L.J., Bustos, P., Núñez, P., Macharet, D.G.: Socially acceptable robot navigation over groups of people. In: IEEE Conference on Robot and Human Interactive Communication, RO-MAN2017 (2017)
5. Charalampous, K., Kostavelis, I., Gasteratos, A.: Robot navigation in large-scale social maps: an action recognition approach. Expert Syst. Appl. **66**, 261–273 (2016)
6. Thrun, S., Gutmann, J.S., Fox, D.: Integrating topological and metric maps for mobile robot navigation: a statistical approach. In: AAAI/IAAI, pp. 989–995 (1998)
7. Kruse, T., Pandey, A.K., Alami, R., Kirsch, A.: Human-aware robot navigation: a survey. Robot. Auton. Syst. **61**(12), 1726–1743 (2013)
8. Kostavelis, I.: Robot behavioral mapping: a representation that consolidates the human-robot coexistence. Robot. Autom. Eng. **1**, 1–3 (2017)
9. Rios-Martinez, J., Spalanzani, A., Laugier, C.: From proxemics theory to socially-aware navigation: a survey. Int. J. Soc. Robot. **7**(2), 137–153 (2015)
10. Charalampous, K., Kostavelis, I., Gasteratos, A.: Recent trends in social aware robot navigation: a survey. Robot. Auton. Syst. **93**, 85–104 (2017)
11. Mumm, J., Mutlu, B.: Human-robot proxemics: physical and psychological distancing in human-robot interaction. In: Proceedings of the 6th International Conference on Human-robot Interaction, pp. 331–338 (2011)

12. Walters, M.L., Oskoei, M.A., Syrdal, D.S., Dautenhahn, K.: A long-term human-robot proxemic study. In: 2011 IEEE on RO-MAN, pp. 137 –142 (2011)
13. Khambhaita, H., Alami, R.: Assessing the social criteria for human-robot collaborative navigation: a comparison of human-aware navigation planners. In: 2017 26th IEEE International Symposium on Robot and Human Interactive Communication (RO-MAN), pp. 1140–1145 (2017)
14. Unhelkar, V., Pérez-D'Arpino, C., Stirling, L., Shah, J.A.: Human-robot co-navigation using anticipatory indicators of human walking motion. In: IEEE International Conference on Robotics and Automation (ICRA), pp. 6183–6190 (2015)
15. Broz, F.: Planning for human-robot interaction: representing time and human intention. Robotics Institute. Carnegie Mellon University (2008)
16. Esteves, C., Arechavaleta, G., Laumond, J.-P.: Motion planning for human-robot interaction in manipulation tasks. In: IEEE International Conference on Mechatronics and Automation, vol. 4, pp. 1766–1771 (2005)
17. Calderita, L.V.: Deep state representation: a unified internal representation for the robotics cognitive architecture CORTEX. University of Extremadura (2015)
18. Beetz, M., Jain, D., Mösenlechner, L., Tenorth, M.: Towards performing everyday manipulation activities. J. Robot. Auton. Syst. **58**(9), 1085–1095 (2010)
19. Manso, L.J., Bustos, P., Bachiller, P., Núñez, P.: A perception-aware architecture for autonomous robots. Int. J. Adv. Robot. Syst. **12**, 174 (2015). https://doi.org/10.5772/61742. ISSN: 1729-8806, InTech
20. McDermott, D., Ghallab, M., Howe, A., Knoblock, C., Ram, A., Veloso, M., Weld, D., Wilkins, D.: PDDL-the planning domain definition language (1998)
21. Haut, M., Manso, L.J., Gallego, D., Paoletti, M., Bustos, P., Bandera, A., Romero-Garces, A.: A navigation agent for mobile manipulators. In: Robot 2015: Second Iberian Robotics Conference, Advances in Intelligent Systems and Computing, vol. 418, pp. 745–756 (2015)
22. Hall, E.: Proxemics. Curr. Anthropol. **9**(2–3), 83–108 (1968)
23. Okal, B., Arras, K.O.: Learning socially normative robot navigation behaviors with Bayesian inverse reinforcement learning. In: 2016 IEEE International Conference on Robotics and Automation (ICRA), pp. 2889–2895 (2016)

Talking with Sentiment: Adaptive Expression Generation Behavior for Social Robots

Igor Rodriguez[1]([✉]), Adriano Manfré[2], Filippo Vella[2], Ignazio Infantino[2], and Elena Lazkano[1]

[1] Computer Sciences and Artificial Intelligence, University of Basque Country (UPV/EHU), Manuel Lardizabal 1, 20018 Donostia, Spain
igor.rodriguez@ehu.eus
[2] Institute for High Performance Computing and Networking (ICAR), National Research Council of Italy (CNR), Palermo, Italy
adriano.manfre@icar.cnr.it

Abstract. This paper presents a neural-based approach for generating natural gesticulation movements for a humanoid robot enriched with other relevant social signals depending on sentiment processing. In particular, we take into account some simple head postures, voice parameters, and eyes colors as expressiveness enhancing elements. A Generative Adversarial Network (GAN) allows the proposed system to extend the variability of basic gesticulation movements while avoiding repetitive and monotonous behavior. Using sentiment analysis on the text that will be pronounced by the robot, we derive a value for emotion valence and coherently choose suitable parameters for the expressive elements. In this way, the robot has an adaptive expression generation during talking. Experiments validate the proposed approach by analyzing the contribution of all the factors to understand the naturalness perception of the robot behavior.

Keywords: Social robotics · Cognitive robotics
Human-robot interaction · Generative adversarial network

1 Introduction

Social robots represent a great research challenge, aiming to an effective introduction in human everyday life of intelligent embodied machines. Robot social capabilities require both a deep understanding of human behavior and acting with naturalness during the interaction with humans [10]. Naturalness means that a human user could have similar perceptive inputs while interacting with other people considering both verbal and non-verbal signals, social and cultural context, subjectiveness and psychological effect as empathy, emotional impact, and so on.

© Springer Nature Switzerland AG 2019
R. Fuentetaja Pizán et al. (Eds.): WAF 2018, AISC 855, pp. 209–223, 2019.
https://doi.org/10.1007/978-3-319-99885-5_15

The gestures, postures, and movements of the body and face expressions are used to convey information about the emotions and thoughts of the sender while supporting verbal communication. Body language represents the key to express feelings and helps the people to understand sociability [14]. McNeill [16] distinguishes four major types of gestures by their relationship to the speech: deictic, iconic, metaphoric, and beats. Unlike the others types, beats are not associated with particular meanings, and they occur with the rhythm of the speech. Such kind of gestures have been considered in this work. While speaking, the robot has to generate credible body language that should shape and convey the information content. It can be derived and learned from humans so that it is consistent with socio-cultural expectation of the interlocutor. Moreover, the robot has to own an emotive model to dynamically drive the interaction and to establish a relevant emotional link with the interlocutor.

The main contribution of this work is the development of a robot behavior that endows humanoid robots with the ability to generate natural gesticulation movements enriched with several social signals depending on the sentiment of the speech. A Generative Adversarial Network (GAN) allows the proposed system to extend the variability of some basic gesticulation movements avoiding repetitive and monotonous behavior. Furthermore, we take into account some simple head postures, voice parameters and eye LEDs colors to enhance the expressiveness of humanoid robots. Two different experiments have been performed with people in front of a SoftBank's Pepper robot showing our adaptive expression generation behavior. Experiments validate the proposed approach by analyzing the contribution of each of the factors to understand the naturalness perception of the robot behavior.

2   Related Work

Social robotics [4] aims to provide robots with artificial social intelligence to improve human-machine interaction and to introduce them in complex human contexts. The demand for robot's sophisticate behaviors requires to model and implement human-like capabilities to sense, to process, and to act/interact naturally by taking into account emotions, intentions, motivations, and other related cognitive functions. In recent years a lot of effort has been put in trying to make those behaviors convey sentiment. Several works propose facial expressions as principal mechanism to show emotions, but there are also other possibles communicative channels that can be easily understood by a human. For example, colors can be dynamically associated with emotions by suitable cognitive models [2,11]. Low-resolution RGB-LEDs can evoke associations to basic emotions (happiness, anger, sadness, and fear), by using suitable colors and dynamic light patterns [7]. Johnson et al. [12] investigate how LED patterns around the eyes of Softbanks NAO robot can be used to imitate human emotions.

As a matter of fact, postures and movements are relevant for social interactions even if they are subjective and culture dependent. During verbal communication, the level of trust of the human with respect to the robot is higher when the robot's gaze is in the direction of the interlocutor [19]. In [1], authors propose a multimodal robot behavior, expressed through speech and gestures, in which the robot adapts its behavior to the interacting human's personality, and they explore the perception of the interacting human comparing the multimodal behavior with the single-modal behavior, expressed only through speech.

In the field of computer graphics and virtual agents, the results obtained on realistic animation of humans are impressive and allow designers to animate characters with movements by synchronizing nonverbal behavior with synthesized speech [5,17]. Naturally, speech plays a relevant role to convey emotions, and human voice can be shaped in a very complex way. In the context of human-robot interaction, Crumpton and Bethel explain the importance of using vocal prosody in robots to convey emotions [6].

With respect to the aim of the present work, to the best of our knowledge, the literature does not provide an approach that tries to combine all the previous aspects in a social robot. In the following, we introduce our approach to generate an adaptive expression behavior for social robots.

3   System Architecture

In this section, the architecture of the expression generation behavior, which we have named "Adaptive Talking Behavior", is described. The expression generation process can be summarized in three main steps:

1. Extract the sentiment from the text. A sentiment analyzer assesses the sentiment of the text and gives as output a descriptor with information about the polarity of the sentiment (positive/negative/neutral).
2. Sentiment to emotion conversion. In this step, the sentiment polarity is encoded into emotion. Only sadness, happiness and neutral emotions have been considered in this work.
3. Generate the appropriate expression. The translation from emotion into expression is performed in this stage. The robot shows an emotional expression by means of body expression (talking gestures), facial expression (eyes lighting), and voice intonation (pitch and speed variation).

Those three steps are further detailed in Sects. 4, 5, and 6, respectively.

This "Adaptive Talking Behavior" is composed by several ROS (Robot Operating System)[1] modules, as illustrated in Fig. 1. The "Sentiment Analyzer" analyzes the sentiment of the text, the "Emotion Selector" converts the sentiment into an emotion, the "Eyes Lighting Controller" manages the eyes color, the "Speech Synthesizer" tunes voice parameters, and the "Gestures Generator" generates the body expression including arms, hands and head.

---

[1] http://wiki.ros.org/.

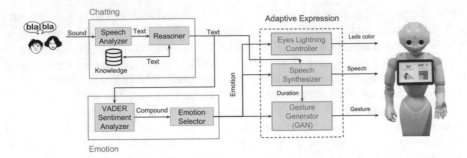

Fig. 1. Description of the Adaptive Talking Behavior architecture

4   Text Sentiment Extraction

Sentiment analysis is the research field related to the analysis of people's opinions, sentiments, evaluations, attitudes, and emotions from written language [18]. The main purpose of sentiment analysis is to extract the polarity (positive/negative/neutral) of a given text.

In order to extract the sentiment from the text we use the VADER sentiment analyzer [9], a lexicon and rule-based sentiment analysis library that analyzes the polarity and the intensity of sentiments expressed in social media contexts, also generally applicable in other domains. VADER, which is based on dimensional affective models, gives an output composed by: (1) the score ratios for proportions of text that fall in each category, and (2) a compound score, obtained by summing the *Valence* scores of each word in the lexicon (see Sect. 5 for a deeper explanation about dimensional affective models).

For the phase described above, we have developed a ROS module, named "Sentiment Analyzer", which takes as input what the robot is going to say (a text) and gives as output the sentiment polarity (negative/neutral/positive) and the compound score obtained by using the VADER sentiment analyzer.

5   Sentiment to Emotion Conversion

Dimensional affective models represent affective experiences according a set of interrelated and ambiguous states [20]. Emotions are described as linear combinations of *Valence-Arousal-Dominance* (VAD). *Valence* defines how positive or negative the stimulus is, *Arousal* specifies the level of energy and *Dominance* defines how approachable the stimulus is.

The *Valence* deals with the positive or negative character of the emotion, which scales from sadness to happiness. Taking into account that the compound score provided by the VADER sentiment analyzer is obtained from the *Valence* scores of the words and then normalized between −1 to +1 (from most negative to most positive), we compute a conversion from sentiment to emotion through a direct translation of the compound score into the sadness-happiness continuum in the *Valence* axis (Happiness: compound score ≥ 0.5, Neutral: compound score > -0.5 and compound score < 0.5, Sadness: compound score ≤ -0.5).

The conversion from sentiment to emotion is done by the "Emotion Selector" ROS module, which takes as input the result obtained from the "Sentiment Analyzer". For the time being, the emotion appraisal is done as a direct translation from the sentiment value into emotion in sadness-happiness continuum. It is worth mentioning that more inputs and more emotions should be considered for the emotion appraisal in the future.

6   Expression Generation

Emotion expression is one of the characteristics that make us social beings. It allows us to communicate our emotional state and, at the same time, it gives us a glimpse into the inner mental state of other individuals. Emotional expressions can occur with or without self-awareness during both verbal and nonverbal communication, and can be manifested in different ways, such as facial movements, body postures, gestures, etc.

Our approach to appropriately express the emotion obtained from the sentiment processing of the text consists of: mapping the emotion into expression that combines natural body gestures, enriched with facial expressions and voice intonation. Several video examples about Pepper talking with sentiment are available in RSAIT's YouTube channel[2].

6.1   Gestures Generation

Typically, humans use gestures (head, hands and arms movements) to communicate with others; gestures are used both to reinforce the meaning of the words and to express feelings through non-verbal signs. How an emotion is reflected in the different parts of the body is well explained in [15]. Humanoid robotics platforms, like Pepper, help us to explore body language expression approaches. Thanks to their high expression capabilities we can build natural robot behaviors that enhance the expressiveness of robots. A previous work by Rodriguez et al. [21], show a robot behavior that makes a NAO robot able to talk and gesticulate showing some emotions. The robot executes randomly selected predefined gestures adapting the execution tempo of the movements according to the sentiment of the text.

The short set of gestures (default animations in the NAOqi API[3]) that Pepper and NAO have for gesticulating while speaking make their expression ability limited and repetitive. In order to overcome this limitation, our approach based on Generative Adversarial Networks (GAN) enables humanoid robots to dynamically generate synthetic gestures (composed by arm, hand and head movements) during verbal communication at run-time.

---

[2] https://www.youtube.com/channel/UCT1s6oS21d8fxFeugxCrjnQ.
[3] http://doc.aldebaran.com/2-5/ref/python-api.html.

Generative Adversarial Networks [8] are semi-supervised emerging models that basically learn how to generate synthetic data from the given training data. A GAN network is composed by two different interconnected networks. The Generator (G) network generates possible candidates so that they are as similar as possible to the training set. The second network, known as Discriminator (D), judges the output of the first network to discriminate whether its input data are "real", namely equal to the input data set, or if they are "fake", that is generated to trick with false data. The general architecture of a GAN with G and the D networks is shown in Fig. 2.

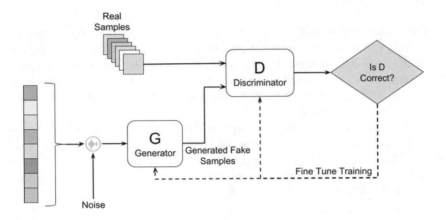

Fig. 2. Description of GAN architecture

In the first step, the D takes as input both, real data and fake data, and returns for each sample its probability to be real or not. In the second step, the G network is trained. While the parameters of D are fixed, in each epoch, the weights of the G network are updated to let the discrimination results on the sample generated by G as near as possible to 1. That is, this second step is aimed to modify the G network in order to be able to generate samples that can trick the D network.

The G network is never exposed to the real data, the only manner to enhance its generation capability is through the interaction with D by means of the output. Instead, D has access to both, real data and fake data, and produces as output the ground truth to know if the data came from the generator or the dataset. The discriminator's output value is exploited by the generator to enhance the quality of the forgeries data.

Regarding to the concrete network structure used in this work, all the layers of both networks are dense, and the neuron activation function is the LeakyReLU. While a standard ReLU implements the function $f(x) = max(0, x)$, the LeakyReLU allows a small gradient when the input is less than zero, as shown in Eq. 1. In our implementation the value of the parameter α has been set to 0.2.

$$f(x) = \begin{cases} x & \text{if } x \geq 0 \\ \alpha x \text{ otherwise} \end{cases} \qquad (1)$$

The generator consists of three dense layers, each of them twice the size of the previous one, followed by another final dense layer whose dimension matches the input dimension and its activation function is tanh. The discriminator follows in some way the inverse architecture, with each layer dimension half the previous one, ending in a final sigmoid neuron. A dropout parameter of 0.3 is also associated to each layer, to diminish the chances of overfitting (Fig. 3).

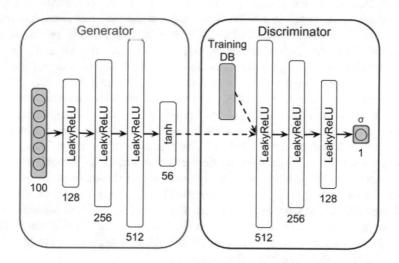

Fig. 3. GAN setup for talking gesture generation

The discriminator network is thus trained using the data to learn its distribution space. The input dimension is 56, corresponding to a vector representing a unit of movement (containing the proprioceptive sensors information of the involved joints). On the other hand, the generator is seeded through a random input with a uniform distribution in the range $[-1, 1]$ and with a dimension of 100. The *Generator* intends to produce as output gestures that belong to the real data distribution and that the *Discriminator* network would not be able to correctly pick out as generated.

After several experiments, we setup a batch size of 16, a learning rate of 0.0002, Adam [13] as the optimization method, and $\beta_1 = 0.5$ and $\beta_2 = 0.999$ as its parameters. We trained the network for 2000 epochs. Back-propagation is applied in both networks to enhance the accuracy of the generator to produce valid movements; on the other side, the discriminator becomes more skilled to false flag data.

The training dataset given to the D network to learn the distribution space of the data is composed by the gestures obtained from the default animations of the NAOqi API. We have chosen a subset of those gestures that can be used for accompanying the speech, and can be performed individually or in composition to constitute complex sequences of gestures. On the other hand, the Generator is seeded through a random input with a uniform distribution in the range $[-1, 1]$. The Generator produces as output gestures that belong to the real data distribution and that the D network is not able to evaluate as generated or real.

In order to extend the variability of the gestures, we sampled the default NAOqi animations with a frequency of 4 Hz, obtaining a set of 1500 robot poses. A pose is represented by a set of 14 joint values, comprising robot's head, hands and arms. The composition of four consecutive poses is a gesture (movement segment) that is one instance of the training set. The GAN network could have been trained using poses instead of movement segments, but then, the outputs would be single poses that should be afterwards concatenated to generate talking movements. However, this approach would produce less smooth gestures.

The talking gestures generation is done by the "Gestures Generator" ROS module. that takes as input the emotion value obtained from the "Emotion Selector" and produces a number of gestures that are well suited to be executed with the right velocity according to the speech duration. The execution velocity is influenced by the recognized emotion in order to express better the feeling, i.e. if the emotion to be shown is "happy" the gesture will be executed at a faster pace than whether to gestures to be performed is bound to the emotion "sad".

Also the head tilt is influenced by the emotion. If the emotion is neutral, the robot will look forward. However, if the emotion is happy the robot will tilt the head upwards, or downwards if it is sad. The emotion *Valence* obtained by the emotion appraisal is normalized between the maximum and minimum values for the head tilt. Figure 4 shows an example of gestures generated using GAN for sadness, neutral and happiness emotions.

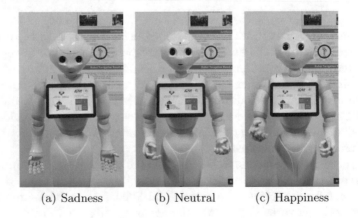

(a) Sadness   (b) Neutral   (c) Happiness

Fig. 4. Some examples of generated gestures with emotion

6.2   Facial Expression

The design of humanoid robots' eyes is usually inspired by human face, trying to exactly reproduce human eyes' shape and movements. However, SoftBank's robots have some limitations due to the structure of their eyes. In particular, Pepper robots' eyes are composed by two rings of LEDs with a black pupil inside. The LEDs can be controlled to show different hues, change color intensity and can be turned on/off for different time duration.

Johnson et al. [12] demonstrate in their work that NAO's eyes can be used to express emotions. Taking inspiration from their color-emotion study, in our approach we adopt the same color configuration, and in addition we use the emotion *Valence* to change the intensity of the color.

The "Eyes Lighting Controller" is employed to convert emotion into facial expression, specifying the color and the intensity of each eye LEDs (see Fig. 5). The controller exploits the emotion *Valence* value to codify it into RGB space to be displayed in the robot.

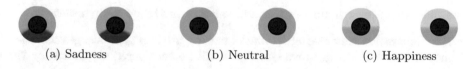

(a) Sadness   (b) Neutral   (c) Happiness

Fig. 5. Sadness: blue-greenish color from RGB(0, 0, 255) to RGB(0, 255, 255). Neutral: no color from last RGB to RGB(0, 0, 0). Happiness: yellow color from RGB(76, 76, 0) to RGB(255, 255, 0).

6.3   Voice Intonation

In ordinary life humans use different voice intonation depending on the context in which they are and also to emphasize the message being conveyed. The voice intonation has a key role to understand the mood of the speaker. The influence of the voice intonation in emotional expression is clearly argued in [3]. The authors prove that some emotions, such as fear, happiness and anger, are portrayed in a higher speech rate and also at a higher pitch than emotions such as sadness.

We have used the happiness, neutral and sadness intonations to portray the three emotions available in our system. Unfortunately, Pepper's speech synthesizer does not offer direct voice intonation selection, but it provides the option to setup voice parameters such as pitch and speech rate, which can be tuned to obtain a different voice intonation than the standard provided. Our approach consists of changing the pitch and speed rate parameter values according to the emotion *Valence* value, i.e. the emotion's *Valence* obtained from the emotion appraisal is normalized between the maximum and minimum values for the voice pitch and speed rate. Maximum and minimum values have been experimentally defined for our system.

7   Experimental Setup

In this section we introduce the robotic platform used in the experiments, the hypotheses we want to validate through the experiments, and an overview for the conducted experiments.

7.1   Robotic Platform

The robotic platform employed in the performed experiments is a Pepper robot developed by Softbank Robotics[4]. Pepper has a height of 120 cm and 20 degrees of freedom human-like torso that is fitted onto a wheeled platform, equipped with full-color RGB LEDs (placed in eyes, ears and shoulders), three cameras, and several sensors located in different parts of its body that allow for perceiving the surrounding environment with high precision and stability.

7.2   Hypothesis

The presented research aims to test and validate the following hypotheses:

- H1: Generative Adversarial Networks can be used to generate gesticulation movements and the gestures generated will be considered as natural by the user.
- H2: The robot expression displayed through generated body gestures adapted and combined with eyes' colors and voice intonation are perceived as more expressive by the user than expressing only through talking arm gestures.

7.3   Experimental Design

In order to test and validate those two hypothesis (H1 and H2), we have defined two different experiments (E1 and E2) in which participants must judge the robot behavior by filling a questionnaire designed to analyze their perception about the system.

- E1: The robot reports a short news (duration of 14 s) extracted from Palermo Today digital newspaper to the participants. It repeats three times the same piece of news using a different type of gesticulation in each session: using *Random* mode of Softbank's Animated Speech module, which launches some neutral animations executed one after another; using *Contextual* mode of Softbank's Animated Speech module, which launches some specific animations each time a keyword is detected, and when no contextual animations is found, it randomly launches a new animation; using gestures generated by GAN.

---

[4] https://www.ald.softbankrobotics.com/en/robots/pepper.

– E2: The robot tells different tweets (7 sec duration approximately) to the participants. It repeats three times the tweets, but adding a new feature to the robot expression each session. First, talking gestures are generated adding head tilt and arms movement speed features (S1). In the next session (S2), the color of the eyes is added together with the features added in S1. Finally, in the last session (S3), the tone of the voice is added together with the features added in S1 and S2.

8 Experimental Results

For the evaluation of the system, 57 voluntaries have been recruited to judge the behavior of Pepper during a talk. The participants grouped into three or four, entered in the experiment room without any information about the experiments and were seated in front of the robot.

In the first experiment (E1), participants evaluated the talking gestures performed by the robot. The order of the type of gestures performed by the robot was randomized in order to avoid possible bias of people always choosing the last remembered as best behavior. After seeing each session participants filled a questionnaire based on five-point Likert scale rating the following aspects: the *naturalness* (A) of the gestures, the *fluency* (B), the *appropriateness* of the gestures for accompanying the speech (C), the *variability* of gestures perceived (D), the *synchronization* (E) between the speech and the gestures, and how much they *liked* (F) the gestures performed by the robot. Additionally, after seeing all sessions, they chose their preferred one.

Table 1. Mean values based on five-point Likert scale rating for each gesticulation type.

	Random	Contextual	GAN
(A) Naturalness	3.4	3.4	3.2
(B) Fluency	3.5	3.6	3.3
(C) Appropriateness	3.2	3.2	3.2
(D) Variability	3.1	3.1	3.0
(E) Synchronization	3.3	3.3	3.3
(F) Liking	3.4	3.4	3.3

Mean score results in Table 1 are very similar for all the gesticulation types. No significant conclusions can be extracted beyond confirming that the GAN approach produces credible movements during the speech even when no contextual information is used. However, when querying the participants about the preferred session GAN based approach obtained 41%, while Random and Contextual obtained 30% and 29% respectively. The main advantage of the

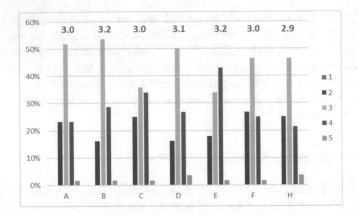

Fig. 6. Results of E2-S1: GAN + Head tilt + Arm velocity. Mean scores are reported above each histogram.

GAN approach is that the robot can use different datasets of simple movements depending on cultural context, social practices, or individual preferences.

In the second experiment (E2) the potentiality of the generated robotic gesticulation (by using GAN) is evaluated considering the possibility to convey also emotions depending on the pronounced speech, and enriching the interaction with other relevant factors such as the head movements and arms movement speed, the tone of the voice, and the color of the eyes. Participants must judge the gestures in the same way as in the first experiment (questions A-F), but they also must identify in which *part of the body* they appreciated the emotion (G), and how much they *liked* the overall robot capability to perform expressions (H).

Table 2. Influence of each relevant part in the different sessions.

(G) Relevant part	S1	S2	S3
Head	79%	64%	71%
Arms	43%	36%	46%
Eyes	2%	80%	70%
Voice	16%	14%	57%

Figures 6, 7, and 8 show the frequency for each point value of the Likert scale rating in each experiment (note that there were no query evaluated as 1). Results in Figs. 6, 7, and 8 show that robot's expressiveness improves significantly by adding more features as eyes LEDs colors and voice tone. In particular, results of Table 2 show the influence of each relevant part in the different sessions: participants clearly appreciate the effect of head tilt when expressing emotions, and they also perceived the influence of changing the arms movement speed according to the emotion; the introduction of the eye LEDs colors during S2 has

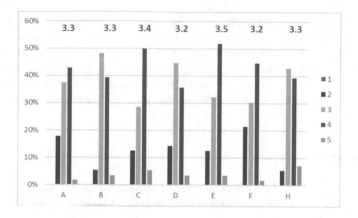

Fig. 7. Results of E2-S2: S1 features + Eye LEDs colors

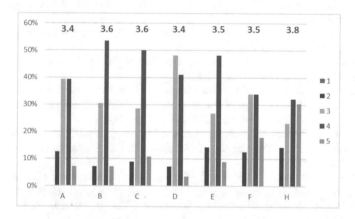

Fig. 8. Results of E2-S2: S1 and S2 features + Voice tone.

a great impact in expressiveness perception; also the modulation of the voice tone is well perceived in S3.

9   Conclusions and Further Work

A suitable generator of rhythmic gesticulation movements with few others expressive features (the head posture, the arm movements velocity, the variation of voice tone, the change of color of eyes) dependent from sentiment detected on sentences, could be a simple system to have high appreciation rates during a human-robot conversation. On the basis of the obtained results, the further work could focus the following interesting directions. The dataset of basic movements used by the GAN approach could be derived from the direct observation of the human by using an RGBD device. In this way, a robot can use with a given person a set of movements that are familiar to him/her. Furthermore, we plan to introduce the detection of the six basic emotions to have a more complex expressive behavior and to consider also some gestures with metaphoric or iconic meaning. We will perform similar experiments exploiting the two different cultural contexts (Italian and Spanish) aiming to investigate for instance the effects of the cultural influences. Moreover, we should investigate to find also evaluation methods to measure the goodness of the gestures generated by the GAN network.

References

1. Aly, A., Tapus, A.: Towards an intelligent system for generating an adapted verbal and nonverbal combined behavior in human-robot interaction. Auton. Robots **40**(2), 193–209 (2016)
2. Augello, A., Infantino, I., Pilato, G., Rizzo, R., Vella, F.: Binding representational spaces of colors and emotions for creativity. Biologically Inspired Cogn. Architectures **5**, 64–71 (2013)
3. Bänziger, T., Scherer, K.R.: The role of intonation in emotional expressions. Speech Commun. **46**(3), 252–267 (2005)
4. Breazeal, C.: Designing Sociable Robots. Intelligent Robotics and Autonomous Agents. MIT Press, Cambridge (2004)
5. Cassell, J., Vilhjálmsson, H.H., Bickmore, T.: Beat: the behavior expression animation toolkit. In: Proceedings of the 28th Annual Conference on Computer Graphics and Interactive Techniques, PP. 477–486. ACM (2001)
6. Crumpton, J., Bethel, C.L.: A survey of using vocal prosody to convey emotion in robot speech. Int. J. Soc. Robot. **8**(2), 271–285 (2016). https://doi.org/10.1007/s12369-015-0329-4
7. Feldmaier, J., Marmat, T., Kuhn, J., Diepold, K.: Evaluation of a RGB-LED-based emotion display for affective agents. arXiv preprint arXiv:1612.07303 (2016)
8. Goodfellow, I., Pouget-Abadie, J., Mirza, M., Xu, B., Warde-Farley, D., Ozair, S., Courville, A., Bengio, Y.: Generative adversarial nets. In: Advances in Neural Information Processing Systems, pp. 2672–2680 (2014)
9. Hutto, C.J., Gilbert, E.: Vader: A parsimonious rule-based model for sentiment analysis of social media text. In: Eighth International AAAI Conference on Weblogs and Social Media (2014)
10. Infantino, I.: Affective human-humanoid interaction through cognitive architecture. In: Zaier, R. (ed.) The Future of Humanoid Robots - Research and Applications. InTech (2012)
11. Infantino, I., Pilato, G., Rizzo, R., Vella, F.: I feel blue: robots and humans sharing color representation for emotional cognitive interaction. In: Biologically Inspired Cognitive Architectures 2012, pp. 161–166. Springer (2013)
12. Johnson, D.O., Cuijpers, R.H., van der Pol, D.: Imitating human emotions with artificial facial expressions. Int. J. Soc. Robot. **5**(4), 503–513 (2013)
13. Kingma, D.P., Ba, J.: Adam: a method for stochastic optimization. arXiv preprint arXiv:1412.6980 (2014)
14. Knight, H.: Eight lessons learned about non-verbal interactions through robot theater. In: Social Robotics, pp. 42–51 (2011)
15. Lhommet, M., Marsella, S.: Expressing emotion through posture and gesture. In: The Oxford Handbook of Affective Computing, pp. 273–285. Oxford University Press (2015)

16. McNeill, D.: Hand and Mind: What Gestures Reveal About Thought. University of Chicago press (1992)
17. Neff, M., Kipp, M., Albrecht, I., Seidel, H.P.: Gesture modeling and animation based on a probabilistic re-creation of speaker style. ACM Trans. Graph **27**(1), 5:1–5:24 (2008)
18. Pang, B., Lee, L., et al.: Opinion mining and sentiment analysis. Found. Trends in Inf. Retrieval **2**(1–2), 1–135 (2008)
19. Paradeda, R.B., Hashemian, M., Rodrigues, R.A., Paiva, A.: How facial expressions and small talk may influence trust in a robot. In: International Conference on Social Robotics, pp. 169–178. Springer (2016)
20. Posner, J., Russell, J.A., Peterson, B.S.: The circumplex model of affect: an integrative approach to affective neuroscience, cognitive development, and psychopathology. Dev. Psychopathol. **17**(3), 715–734 (2005)
21. Rodriguez, I., Martínez-Otzeta, J.M., Lazkano, E., Ruiz, T.: Adaptive emotional chatting behavior to increase the sociability of robots. In: International Conference on Social Robotics, pp. 666–675. Springer (2017)

Mobile Robots

Auction Model for Transport Order Assignment in AGV Systems

Daniel Rivas$^{(\boxtimes)}$ (iD), Joan Jiménez-Jané, and Lluís Ribas-Xirgo (iD)

School of Engineering, Universitat Autònoma de Barcelona, carrer de les Sitges,
Campus UAB, 08193 Cerdanyola del Vallès, Bellaterra, Spain
{Daniel.Rivas,Lluis.Ribas}@uab.cat

Abstract. Systems of automated guided vehicles (AGV) often support internal transportation in flexible manufacturing facilities and warehouses. One of the critical aspects for the efficiency of these systems is the transport order (TO) assignment. Typically, AGVs bid for TOs so that the behavior of the transportation system is affected by the auction type and model parameters, and by AGV estimates on transport costs.

In this work, we have adapted an auction model from a cab service in a city. TOs include a picking and a destination place. We have added to the basic auction model an iterative process to account for any system changes while assigned taxis approach their corresponding picking places. It has been validated in a simulator, which can be used to tune the model parameters in accordance with a given scenario (i.e. floorplan, number and type of AGVs, TO sequence profiles, *et cetera*) and, obviously, for automating the process of TO assignment.

Keywords: AGV · Internal logistics · Transport orders management
Auction

1 Introduction

Mass production of heavily customized products to be delivered just in time is only possible through automation, including that of internal transportation. Autonomous transportation inside industrial plants and the like has, thus, become commonplace, as it is the only way to keep the pace of the industry needs.

Automated guided vehicles (AGV) take care of goods transportation, cleaning and other tasks [1–7] in an autonomous form while being more robust, flexible and adaptable than other solutions. In fact, AGV systems have the capability to deal with the changing characteristics of the dynamic environment that is present in many industry and services' facilities [8–10].

As to have an idea of the market penetration of AGV, in 2015, there has been a yearly worldwide production of 14,000 AGVs [11]. On the next year, the AGV market was estimated at USD 1.12 billion [12], and it is expected to double that number before 2022 [13, 14].

One of the challenges for automated internal transportation is that of assigning a transport order (TO) to a given AGV.

© Springer Nature Switzerland AG 2019
R. Fuentetaja Pizán et al. (Eds.): WAF 2018, AISC 855, pp. 227–241, 2019.
https://doi.org/10.1007/978-3-319-99885-5_16

In this regard, TO management systems for outdoor transportation systems [15–17] are not particularly fit for internal logistics. Indoor transportation takes place in a more structured environment, with less incidences and operation tolerance, e.g. service times must be met within shorter variation ranges, and agent communications must be agile and quick.

Yet, we address this problem by solving an analogous one: assign clients to taxis in a city. For that, we took a simple, working base model called *taxisCooperative*, which deals with the transport of passengers from a random origin to a fixed destination, an airport, in the original case [18]. This model already implements a basic auction solution between the agents to decide which taxi does the transportation (Sect. 3).

In this work, we adapted the *taxisCooperative* model to emulate a set of taxis taking clients to different locations, which resembles a system of transportation vehicles in a warehouse or a factory, and we proposed a new auction mechanism to obtain a more efficient transportation. The efficiency depends on total costs per transportation, and cost estimation is essential for bidding in TO auctions (Sect. 2).

For the sake of simplicity, we have modified the base model preserving its original function, i.e. that of picking clients at their places and dropping them at their destinations (Sect. 4). The analogy to AGVs doing transportation in an industrial facility is obvious.

The new model has been validated by simulating a range of cases with varying number of taxis and TOs (Sect. 5). The addition of secondary auctions has shortened the service times significantly.

2   Transport Cost Estimation

TOs are auctioned among AGV agents, which estimate transport costs to place a bid. As already said, one of the main differences between outdoor and indoor transportation is on the accuracy of the estimates.

Eligible bidders are AGVs that can reach the picking spot within the maximum allotted time and have energy enough to fulfill the corresponding TO. Typically, the winner is the one that will incur in the smallest expenditure of resources within the eligible ones, i.e., the one with the shortest time to job completion and with the most efficient energy usage [19].

To estimate the cost of a specific transport, the path from the current location to the destination via the picking place must be computed. In recent years there has been an extensive research on path planning methods and algorithms to direct the navigation of mobile robots. Such path planners may use different approaches for the robots' navigation. They may take into account the presence of human beings [20, 21], may be sensitive to the general context [22, 23] and even create their own maps of the operation area [22–24].

The time to job completion can be indirectly determined by the distance of each AGV to the target location. Nevertheless, it is much accurate if previous travel times for the same path are used [25]. Differences in path choices can be significant and so the result of the TO assignment.

Historic data on travel times for path segments can reveal many factors of AGVs and environment that can be extracted to be shared with other AGVs or to tune system-wide model parameters. Particularly, the energy level of AGV batteries and the floor conditions. Evolution of these variables can be tracked down for each AGV and used to improve travel time estimates [26, 27].

In some situations, with work areas shared with humans or with lots of vehicles in a large, unrestricted movement area, AGV reliability and human behavior predictability are also significant factors. These and other factors add complexity to the cost estimation problem, which has triggered the development of many task assignment methods with different approaches to find the optimal agent to perform a specific TO [28–31].

Independently of the factors, formula and method to compute the path cost, its value determines that of the bid of an AGV for a TO. However, once an AGV wins the bid and the TO allocation is done, none of the reviewed methods checks again for a possible better solution that may appear after the allocation procedure's conclusion. For that reason, we believe that our second auction model may coexist with those methods and improve their results in terms of system's robustness, adaptability and efficiency.

In this work we shall focus on the auction mechanism rather than on the cost estimation procedure. With no loss of generality, since our approach does not directly improve the cost estimation method, we used path length as cost because other techniques require measuring real data. For all that, a comparison of the developed model with other existing ones is not suitable except for average service times, however they are measured.

3 The *taxisCooperative* Model

The *taxisCooperative* model is a multi-agent system model of a set of taxis taking clients at random positions to the airport [18]. Passenger agents initiate Contract Nets to get the nearest taxi. The simulation of the system is done in two steps: the configuration and the execution.

The configuration step creates the simulation scenario, including the map and the taxis. As for the map, it is divided in cells and each cell can either be part of a street or of a non-circulating spot (rectangular city building blocks or the airport). Street cells are tagged with the distance to the airport (*distance-airport* variable) and classified whether they are in a crossing or not.

During the execution phase, while they are not picking up or transporting a passenger, the taxis navigate the map in a randomly and autonomous way. The passengers appear on the scene at random positions and time. This means that all the passengers do not appear at the same time, but they do it progressively, with a probability of 30% at each simulation step. When a passenger appears on the stage, an auction begins to decide which taxi will carry out the transportation. Therefore, each passenger is equivalent to a transportation order.

The type of auction applied in the model is a *closed envelope auction*, in which all bidders submit their offer to the auctioneer once and secretly. Therefore, none of the bidders knows the price that the others have presented. Once the auctioneer has all the offers, it checks them and chooses the one with the highest price as the winner. In the case of the *taxisCooperative* model, the winner is not the taxi with the highest cost, but the one with the lowest cost. The reason behind this is that a higher cost implies that the taxi is at a worse position to carry out the transport operation, meaning that it will spend more time and energy to reach the target location.

When a passenger appears, it sends a message to all the available taxis, requesting one to go to the airport, along with the coordinates where the passenger is located. When the taxis receive the message, they calculate the distance in a straight line from their location to the passenger's location, as depicted in Fig. 1, and communicate it back to the passenger. The passenger will collect the taxis answers during 15 simulation cycles, and once this period is over, it will evaluate all the collected messages and will choose the one with the lowest value. At this point the auction has ended and the passenger communicates to the winning taxi to pick it up.

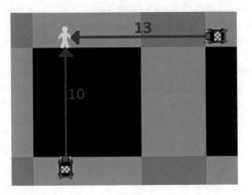

Fig. 1. Cost calculation.

The taxi finds the junction that is closest to the passenger using the Euclidean distance. Then, it sets course towards the coordinates of the junction, and once there, it sets course towards the passenger's coordinates.

Once the taxi has picked up the passenger, it will take it to the airport. To take the shortest route to go to the airport, the taxi looks at the *distance-airport* value of the streets and junctions on its surroundings and goes towards the one with the lowest value, this procedure is repeated until it arrives to the airport and leaves the passenger. Once they have arrived at the airport, the taxis leave the passengers and are free again to make a new trip.

Figure 2 shows how the service time of a taxi is structured and the elements that compose it. First, a passenger appears and requests a taxi, which marks the beginning of the auction period. Once the auction is finished, the winner taxi starts its travel to the passenger's location, starting the pick-up stage. When the taxi arrives at the passenger's location and picks it up, the pick-up time phase ends and begins the delivery phase, which ends when the taxi arrives to the airport and leaves the passenger.

Fig. 2. Service times scheme for a taxi.

4 Modified *taxisCooperative* Model

The *taxisCooperative* model, though realistic, is far from reflecting what occurs in an internal logistics environment. The four main differences are:

- All transportations finish at the same location, the airport. This does not correspond to an internal logistics situation, where each TO may contain different origin and destination points.
- The method for calculating the distance between the taxis and the passenger is too simple and not close to the reality. Each taxi calculates the Euclidean distance to the passenger, rather than the actual distance it would have to travel to go to the passenger's location (avoiding the non-circulating zones). This method does not ensure that the winning taxi is the closest one to the passenger. In Fig. 1 can be seen how this calculation is not the most accurate.
- The different streets and junctions have a variable called *airport-distance* in which the distance from the street/junction to the airport is saved during the configuration phase. When going to the airport, the taxis check all the street's segments in their surroundings, looking for the smallest value of this variable to determine the shortest path to the airport. This routing method can only be used in a scenario where there is only one destination point.

In this section we shall explain how we addressed each one of these issues to make the model closer to the case of internal logistics, and how we added new functionalities to improve its efficiency.

4.1 Remove the Airport from the Map

The airport was removed from the map in order to have a model with several destination points for the TO, rather than just one, which is closer to the reality of internal logistics. As a result, some modifications have had to be made to the model for it to work properly:

- The streets and crossings have been modified so that they no longer have the *airport-distance* variable.
- The passengers' behavior has been modified. Now when they appear in the map and ask for a taxi, they must communicate their location as well as the destination coordinates. All these coordinates are generated randomly.
- When a taxi picks up a passenger, it must calculate the shortest route to reach the destination point.

4.2   Change the Minimum Route Calculation Method

Previously, the taxis used the Euclidean distance from their own locations to the passengers' locations to bid in the transportation auction. This calculation method has been changed by the Dijkstra algorithm [32], which will always give the minimum cost path to reach the passenger taking into account the presence of not circulating areas. This change ensures that the winning taxi is the one that will have to travel the shortest distance to reach the passenger's location.

4.3   Changes in the Map Creation Method

A function to generate an equivalent topological graph is added to the creation method of the map, as seen in Fig. 3, in order to apply the Dijkstra algorithm. All the calculations that were made during the configuration phase, to determine the distance from each street cell to the airport, have been replaced by the interpretation of a matrix representing the work area map. This matrix contains all the elements that should appear in the model, that is: roads, junctions, houses, non-circulating zones and the nodes of the topological graph.

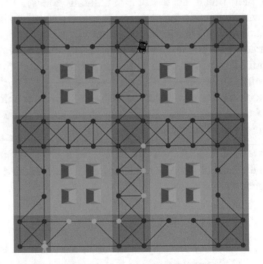

Fig. 3. Topological graph of the model.

It is important to say that the nodes are created but not the arcs that relate them. The drawing of the arcs between the nodes is done passing 3 times by the nodes, each time the algorithm tries to make a different type of arc:

- First time: The algorithm checks node by node if it can establish a horizontal arc towards another node, without going through a non-circulating zone.
- Second time: The algorithm checks node by node if it can establish a vertical arc towards another node, without going through a non-circulating zone.

- Third time: The algorithm tries to establish diagonal arcs towards other nodes, without going through a non-circulating zone. The diagonal that is drawn is that of a square, if the other nodes are outside of this range they are not linked.

4.4   Change the Auction Duration

As explained in Sect. 3, the original auction lasts 15 program cycles. This entails a problem, 15 cycles have passed since the taxi has calculated the best route to go for the passenger until the end of the auction, and therefore the taxi could be far from the initial node. To solve this issue, auctions would be done in a single program cycle, i.e. the duration of the auctions was set to one.

4.5   Second Auction Concept Application

The second auction system establishes that the passenger, once the auction is done and the closest taxi is on its way to pick it up, keeps making auctions to determine if there is a closer taxi. In the case that another taxi is closer, the TO changes to that one and from that moment that new taxi will make the transportation.

The process does not end there, the system continues making second auctions until the cost of the winning taxi is lower than the 20% of the cost calculated by the winning taxi on the first auction. This percentage has been determined after making several executions and analyzing the resulting data. These data will be seen in more detail in Sect. 5. The reason to stop making second auctions at this point is because it has been considered that the winning taxi is close enough to the pick-up point and because from the 20% of the original cost the number of taxi changes is very little.

The following example illustrates better the second auction concept. Let's suppose there are 2 taxis (T_1, T_2) and 2 passengers (P_1, P_2). The initial scenario can be seen in the left part of Fig. 4. P_1 has requested a taxi and the auction winner is T_1 since T_2 is transporting P_2 to its destination. At that time, T_1 goes to pick up P_1 and T_2 is transporting P_2. When T_2 arrives at the destination point and leaves P_2, it will be closer to P_1 than T_1. In this situation, the next second auction performed by P_1 will select T_2 as the winner and assign the transportation task to it, making a taxi change. The right part of Fig. 4 shows the taxi change.

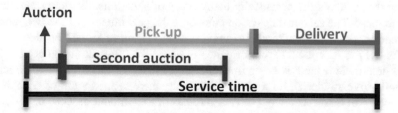

Fig. 4. Service time scheme for a taxi after the modifications in the model.

Finally, in Fig. 5 it can be seen a scheme of the taxis' service time in the modified model. This time the scheme is more complex than the original since the second auction factor is included. The scheme contains the same elements of the service time in the original model but includes the second auction period which works concurrently with the pick-up time until the taxi is relatively close to the pick-up point.

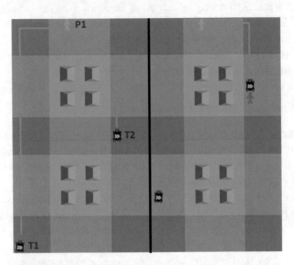

Fig. 5. Second auction example. Left side: T_1 gets the transportation order from P_1. Right side: After delivering P_2 to its destination, T_2 is closer than T_1 to P_1 and receives the transportation order from P_1.

5   Results

In this section will be discussed the results obtained in this work and their impact. First, an analysis on the number of taxi changes that occur at different segments of the routes from the taxis' locations to the pick-up points is performed to determine when the second auction repetitions should stop. The analysis is done with data obtained from a total of 169 combinations of different TOs and taxis amounts, which allowed to evaluate the system's performance in many types of situations, like taxi saturation and taxi emptiness. The second subsection compares the performance of the two models to determine whether the modifications applied had a positive impact on the system or not. Finally, the new model's performance is tested for different amounts of TOs and taxis to determine if there is an optimal amount of taxis for this model and which is it. This could give an idea if, for a given model size, there is an amount of taxis that will always perform better than any other, regardless of the amount of TOs; or if there is a closer relation between the amounts of TOs and taxis.

5.1   Second Auction Analysis

This subsection contains an analysis of the second auction to determine the threshold to stop performing it. On the other hand, it will be seen how the number of taxi changes behave at each point of the route.

In order to collect the necessary data, a group of executions was carried out where, for each taxi and for each TO, the percentage of the taxi changes was captured. These executions were carried out with the threshold value at 0, and therefore there could be taxi changes during the whole route. The executions have been made with different configuration parameters, ranging from 1 to 60 taxis, and from 1 to 60 passengers, increasing these values by 5 in each different execution.

It should be clarified that the appearance of the passengers is done progressively, following a random distribution with a 30% probability of appearance in each simulation cycle. The random distribution causes that the appearance of the passengers in the map does not follow a defined pattern, like what happens in an industry that manufactures on demand. The 30% probability of appearance ensures that the passengers appear progressively during the simulation and not all at once or one right after the other.

Of course, this representation does not include all the possible scenarios that can be found in internal logistics problems. Many working areas have well defined pick-up and delivery points, which would result in a lesser number of possible routes than the ones used in the model. Also, the TOs concentration given by the passengers' probability of appearance does not fit all possible scenarios since this concentration may vary for each application and it also depends on the demand. All this means that the results shown may vary for cases where the probability of occurrence is different. However, the results achieved should not be far from the ones that could be obtained in other internal logistics situations, since the main elements of this kind of scenarios where maintained.

The data from the different executions have been normalized to group them all. Therefore, when interpreting the data, it will not be in meters or kilometers, but in percentages. In the graph, in Fig. 6, it is quantified the times that there has been a taxi change in a distance interval of the established route. This illustrates which segments of the route are the ones with more taxi changes as well as the segments with less changes. The [0, 5] interval refers to changes occurring when the taxi is between 0 and 5% of the initially calculated distance to the destination. The >100 interval means that there has been a change of taxi at a distance farther than the route origin.

Taxis saturation have a great impact in the number of taxi changes. The taxis are programmed to apply a series of techniques to avoid collisions between them. So, if the system is saturated with taxis there will be so many taxi encounters and evasion maneuvers that the taxis will not able to complete their assignments properly and the system will collapse. This phenomenon causes that the interval with most of the taxi changes is the one from 0 to 5% of the route, with more than 450 changes. When a taxi is near the pick-up point and it encounters another taxi, its evasion actions may cause the second taxi to be slightly closer to the pick-up location and a taxi change occurs. However, since the first taxi is already going on that direction more encounters could happen that would trigger more taxi changes.

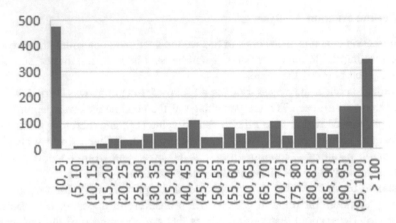

Fig. 6. Number of taxi changes during the second auction (Y axis) against the % of the route where they were made (X axis).

In the interval >100% there are more than 300 taxi changes, which is also caused by the taxis saturation. The collision avoidance techniques can move the taxis away from the original route's plan and from the objective. In the event that a taxi moves away from the destination, there is the possibility that another taxi results to be closer and, therefore, that a change is made.

It must be considered that the second auction must be stopped in a reasonable interval: it should not be repeated when the taxi is very close to the passenger's location, nor when it is very far from it. If the threshold is very close to the passenger's location, it can have a counterproductive effect. If a taxi is about to arrive to the pick-up point and the order changes to another taxi, they may collide and therefore the service time could be affected. It is also not good to have a threshold very early in the pick-up period, since it would be limiting too much the possibilities of taxi changes, taking away the advantage of the second auction. A balance must be sought between the distance and the volume of changes made in each interval.

From the results shown in Fig. 6, the [5, 10] interval contains 0 taxi changes, but it is very close to the destination, so it is not considered a valid threshold. The intervals that go from [10, 15] to [25, 30], have few taxi changes and are at reasonable distances from the destination, so they are potential candidates to be used as the threshold. In this case, the threshold has been set at the 20% since it is the middle value in this low-taxi-exchanges region and, also, at a reasonable distance from the destination.

5.2 Comparison Between the Base Model and the Modified One

This section offers a comparison between the amounts of cycles needed by each model to complete the simulations performed.

It does not make much sense to compare the execution times of the two models given the randomness that exists in terms of the TOs appearance. That is, it is not possible to simulate the same scenarios in the two models. Nor is it entirely correct, because in a real case it does not interest the execution time, but the service time of the taxis. What does make sense is to compare the messages exchanged between the agents, the total traveled distance with one method and the other, the number of collisions that occur and the service times. All these data have not been obtained, and therefore, the models cannot be specifically compared in these ways.

However, when comparing the number of cycles that each simulation lasts, the service times of the adapted model are lower than the ones in the base model. This is depicted in Fig. 7, where the Y axis shows the diminution percentage of the number of cycles in the modified model respect to the base model. This percentage was calculated using Eq. (1) where DP is the diminution percentage, N_{MOD} is the number of cycles used by the modified model and N_{BAS} is the number of cycles used by the base model.

$$DP = 100 - (100 * N_{MOD}/N_{BAS})$$ (1)

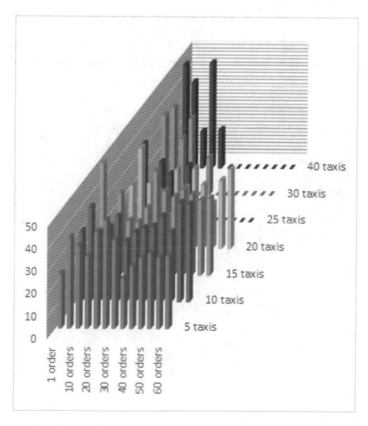

Fig. 7. Diminution of the number of cycles per simulation (in percentage) of the modified model respect the base model (Y axis), against the number of TOs (X axis), and the number of taxis (Z axis), used in the simulations.

Figure 7 also shows that when the amount of TOs was between 30 and 60, and the amount of taxis was high, for the chosen model size, the adapted model does not lower the number of cycles that the base model takes to carry out the TOs. This phenomenon is because the adapted model does not always end correctly when there are too many taxis in the map. When the number of taxis in the model is high, the second auction communications collapse, causing the simulation to go on indefinitely. In contrast, the base model works correctly in these situations and the executions are finished correctly. In the other combinations, the cycles taken by the adapted model fall between 20 and 50%, depending on the case, with respect to the base model. This issue in the modified model is related to a problem in the implementation and will be fixed in future versions.

5.3   Optimal Configuration

This section shows which is the most efficient number of taxis for a given number of TOs, in the modified model.

Figure 8 shows the number of cycles each execution needed to accomplish all TOs, for different amounts of taxis and TOs. These results show that, for the model size that has been used, regardless of the number of TOs, there is not a significant improvement on the performance when there are more than 15 taxis in the system. Actually, if the

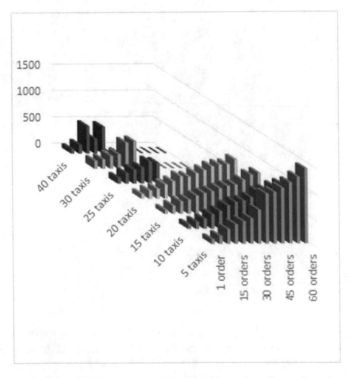

Fig. 8. Average number of cycles per execution (Y axis), against the number of transport orders (X axis), and the number of available taxis (Z axis).

amount of taxis is too large, the system collapses and cannot resolve the assignments. It can be deduced that 15 taxis can address all incoming TOs quickly enough without saturating the model, where a collapse could occur.

6 Conclusions

Transport order assignment is a key element for transportation systems' efficiency and, as systems' complexity grow, it can only be done automatically. In this work, we have explored how this automation can be done in the internal logistics arena.

The main objective of this work was to develop an auction model for TOs using a multi-agent architecture. The model allows progressively incoming TOs to be carried out by a set of taxis. The basic model assigns TOs to taxis by means of a "closed envelope" type of auction. On top of that, we have proposed a repetition of auctions to reduce the system's average service time.

The concept of the "second auction" has been applied, turning the base auction model into one more flexible in terms of taxi availability. This ensures an auction mechanism that adapts to dynamic systems with bidders that may appear in later stages and incidents that may affect the service.

The new model reduces the average service time, measured in simulation cycles, between 20 and 50% as compared to the base model, fulfilling the expectations we had on taking advantage on the system's dynamism. However, the amount of improvement in a real case might be smaller since, in real cases, idle transport agents do not move randomly. This kind of cases will be studied in future works.

Also, in the near future, we shall formalize this model so that initial guesses of its parameters (number of taxis, second auction stop threshold) for a given case can be obtained beforehand and simulations can be used for fine-tuning.

References

1. Franke, H., Dangelmaier, W.: Decentralized management for transportation-logistics: a multi agent based approach. Integr. Comput. Eng. **10**(2), 203–210 (2003)
2. Gehrke, J.D., Herzog, O., Langer, H., Malaka, R., Porzel, R., Warden, T.: An agent-based approach to autonomous logistic processes. KI - Künstliche Intelligenz **24**(2), 137–141 (2010)
3. Hallenborg, K.: Decentralized scheduling of baggage handling using multi-agent technologies. In: Levner, E. (ed.) Multiprocessor Scheduling: Theory and Applications, pp. 436–460. Itech Education and Publishing, Vienna (2007)
4. Tarau, A.N., De Schutter, B., Hellendoorn, H.: Model-based control for route choice in automated baggage handling systems. IEEE Trans. Syst. Man Cybern. Part C (Appl. Rev.) **40**(3), 341–351 (2010)
5. Ribas-Xirgo, Ll., Miro-Vicente, A., Chaile, I.F., Velasco-Gonzalez, A.J.: Multi-agent model of a sample transport system for modular in-vitro diagnostics laboratories. In: Proceedings of 2012 IEEE 17th International Conference on Emerging Technologies & Factory Automation (ETFA 2012), pp. 1–8 (2012)

6. Ribas-Xirgo, Ll., Chaile, I.F.: Multi-agent-based controller architecture for AGV systems. In: 2013 IEEE 18th Conference on Emerging Technologies & Factory Automation (ETFA), pp. 1–4 (2013)
7. Davidsson, P., Henesey, L., Ramstedt, L., Törnquist, J., Wernstedt, F.: An analysis of agent-based approaches to transport logistics. Transp. Res. Part C Emerg. Technol. 13(4), 255–271 (2005)
8. Himoff, J., Rzevski, G., Skobelev, P.: Magenta technology multi-agent logistics i-Scheduler for road transportation. In: Proceedings of the fifth international joint conference on Autonomous agents and multiagent systems - AAMAS 2006, p. 1514 (2006)
9. Ribas-Xirgo, Ll., Chaile, I.F.: An agent-based model of autonomous automated-guided vehicles for internal transportation in automated laboratories. In: ICAART 2013 – Proceedings of 5th International Conference on Agents and Artificial Intelligence, vol. 1, pp. 262–268 (2013)
10. Santa-Eulalia, L.A., Halladjian, G., D'Amours, S., Frayret, J.-M.: Integrated methodological frameworks for modelling agent-based advanced supply chain planning systems: a systematic literature review. J. Ind. Eng. Manag. 4(4), 624–668 (2011)
11. NDC Solutions: The AGV market is booming in silence. https://ndcsolutions.com/insights-news/articles/the-agv-market-is-booming-in-silence/. Accessed 23 Apr 2018
12. Grand View Research: Automated Guided Vehicle Market Size, Share & Trends Analysis Report And Segment Forecasts, 2018–2024 (2018). https://www.grandviewresearch.com/industry-analysis/automated-guided-vehicle-agv-market. Accessed 23 Apr 2018
13. Automated Guided Vehicle Market – Global Forecast to 2022 (2016). https://www.marketresearchreportstore.com/shop/automated-guided-vehicle-market-by-type-unit-load-carrier-tow-vehicle-pallet-truck-assembly-line-vehicle-navigation-technology-laser-magnetic-inductive-optical-tape-battery-type-industry. Accessed 18 July 2018
14. Markets and Markets Research: Automated Guided Vehicle Market - Global Forecast to 2022 (2017). https://www.marketsandmarkets.com/Market-Reports/automated-guided-vehicle-market-27462395.html. Accessed 23 Apr 2018
15. Chen, Z.-L., Pundoor, G.: Order assignment and scheduling in a supply chain. Oper. Res. 54(3), 555–572 (2006)
16. Zegordi, S.H., Beheshti Nia, M.A.: Integrating production and transportation scheduling in a two-stage supply chain considering order assignment. Int. J. Adv. Manuf. Technol. 44(9-10), 928–939 (2009)
17. Woo, H.S., Saghiri, S.: Order assignment considering buyer, third-party logistics provider, and suppliers. Int. J. Prod. Econ. 130(2), 144–152 (2011)
18. Sakellariou, I., Kefalas, P., Stamatopoulou, I.: MAS coursework design in NetLogo. In: Proceedings of the International Workshop on the Educational Uses of Multi-Agent Systems (EDUMAS 2009), pp. 47–54 (2009)
19. Kawakami, T., Takata, S.: Battery life cycle management for automatic guided vehicle systems. In: Design for Innovative Value Towards a Sustainable Society, pp. 403–408. Springer, Dordrecht (2012)
20. Sisbot, E.A., Marin-Urias, L.F., Alami, R., Simeon, T.: A human aware mobile robot motion planner. IEEE Trans. Robot. 23(5), 874–883 (2007)
21. Pol, R.S., Murugan, M.: A review on indoor human aware autonomous mobile robot navigation through a dynamic environment. Survey of different path planning algorithm and methods. In: 2015 International Conference on Industrial Instrumentation and Control (ICIC), pp. 1339–1344 (2015)
22. Lu, D.V., Hershberger, D., Smart, W.D.: Layered costmaps for context-sensitive navigation. In: 2014 IEEE/RSJ International Conference on Intelligent Robots and Systems, pp. 709–715 (2014)

23. Cosgun, A., Christensen, H.: Context Aware robot navigation using interactively built semantic maps. Comput. Res. Repos., October 2017
24. Beinschob, P., Meyer, M., Reinke, C., Digani, V., Secchi, C., Sabattini, L.: Semi-automated map creation for fast deployment of AGV fleets in modern logistics. Rob. Auton. Syst. **87**, 281–295 (2017)
25. Das, P., Ribas-Xirgo, Ll.: Adaptive multi-robot control through on-line parameter identification at system level. Universitat Autònoma de Barcelona (2018)
26. Yan, R., Dunnett, S.J., Jackson, L.M.: Reliability modelling of automated guided vehicles by fault tree analysis. In: 5th Student Conference on Operational Research, p. 10 (2016)
27. Yan, R., Jackson, L.M., Dunnett, S.J.: Automated guided vehicle mission reliability modelling using a combined fault tree and Petri net approach. Int. J. Adv. Manuf. Technol. **92**(5–8), 1825–1837 (2017)
28. Nunes, E., Manner, M., Mitiche, H., Gini, M.: A taxonomy for task allocation problems with temporal and ordering constraints. Rob. Auton. Syst. **90**, 55–70 (2017)
29. Kim, Y., Matson, E.T.: A realistic decision making for task allocation in heterogeneous multi-agent systems. Procedia Comput. Sci. **94**, 386–391 (2016)
30. Liu, H., Zhang, P., Hu, B., Moore, P.: A novel approach to task assignment in a cooperative multi-agent design system. Appl. Intell. **43**(1), 162–175 (2015)
31. Choi, H.-L., Brunet, L., How, J.P.: Consensus-based decentralized auctions for robust task allocation. IEEE Trans. Robot. **25**(4), 912–926 (2009)
32. Cormen, T.H., Leiserson, C.E., Rivest, R.L.: Introduction to Algorithms. MIT Press (1990)

Accountability in Mobile Service Robots

Ángel Manuel Guerrero-Higueras[1(✉)], Francisco J. Rodríguez-Lera[2],
Francisco Martín-Rico[3], Jesús Balsa-Comerón[4], and Vicente Matellán-Olivera[5]

[1] Research Institute on Applied Sciences in Cybersecurity (RIASC),
Universidad de León, Av. de los Jesuitas s/n, 24008 León, Spain
am.guerrero@unileon.es
[2] Computer Science and Communications Research Unit (CSC),
University of Luxembourg, Luxembourg City, Luxembourg
[3] Robotics Lab, Universidad Rey Juan Carlos, Madrid, Spain
[4] Robotics Group, Department of Mechanical, Computer and Aerospace Engineering,
University of León, 24071 León, Spain
[5] Supercomputación de Castilla y León (SCAYLE), 24071 León, Spain

Abstract. Service robots sometimes behave in unexpected ways and
may put economic interests or human safety in risk. This can be accepted
in research environments, but it is not going to be tolerated in everyday
use of robots. In addition, regulations for the deployment of autonomous
robots (from home assistants to autonomous cars) are increasing. These
regulations will require at some point systems that could be audited and
that implement facilities for forensic analysis. In this paper, we propose
that these systems have to be integrated in the development frameworks
of robotics software as a mandatory component. We present two design
alternatives for the *de facto* standard for service robotics (ROS: Robotic
Operating System) to enforce safety and security rules based on a customizable black-box-like component.

Keywords: Safety · Security · Autonomous · Black-box
Accountability · Forensic

1 Introduction

Accountability of software-based autonomous systems is not a new challenge. It
has been widely faced in the software industry. For instance, automatic credit
reporting is heavily regulated and different laws have created to guarantee people's right to know the logic involved in automatic decision-making, for example,
about creditworthiness. The same rights about the action took by autonomous
robots will be regulated.

We need precise regulation to certify, explain, and audit robotic systems [1].
For instance, the European Parliament established a set of recommendations
focused on Civil Law Rules on Robotics in January 2017 [2]. These recommendations overview liability elements involved when a robot is used in a public

© Springer Nature Switzerland AG 2019
R. Fuentetaja Pizán et al. (Eds.): WAF 2018, AISC 855, pp. 242–254, 2019.
https://doi.org/10.1007/978-3-319-99885-5_17

space and also consider several uses. It is a good start point to define how to face the problem but still presents too many gaps to be solved at legal level [3].

From our point of view, one of the more relevant aspects of robotics accountability would be the capability of explaining how decisions taken autonomously by robots were made. It is being discussed if this "right for explanation" is already required by existing legislation, for instance, by the european General Data Protection Regulation (GDPR) [6]. In this legislation individuals are guaranteed meaningful information about how automated decision making is made. However, GDPR does not define the scope of information that should be provided. For this reason, we also propose that the accountability mechanisms integrated in development frameworks should be customizable regarding the scope of information to be preserved in the decision making.

In the future, regulation will become binding, therefore should take effect on the robots. Directives must be incorporated into robots deployed in EU countries. However, it is still not envisioned how the commission will monitor whether EU laws are applied correctly, and on time and it should takes action or not over robots.

Some works can be found about verifiability (see [21]) but there are not many references on implementing accountability in autonomous robots. Most of them are more related to the ethical implications, rather than to the real implementation of accountability systems. Existing systems can be mainly classified into two groups: logging systems, as the ones used in robotic competitions, and the ones related to the implementation of ethical rules.

Robotic competitions have become very popular, both as research and as educational tools. From the point of view of research, reproducibility is a key factor and robots entering competitions are required to record data gathered in the competitions. For example, RoboCup or RoCKIn competitions have developed tools to implement this logging. Something similar happens in educational competitions, in this case to perform adequate evaluations of the results. Unfortunately, these tools are usually implemented as external "referees", that is, programs running in off-board computers. In this way, logging has no performance problems (log files have no limits in their size, specific CPUs are used, etc.), robots in the end are not really autonomous.

The implementation of ethical rules in robots has been a recurrent issue. It has been addressed from the sci-fi movies to the legislation on the military use of autonomous weapons. Some solutions have been proposed, many of them taking into account the accountability of these systems. For instance, Arkin in [7] defines an architecture for embedding ethics in a hybrid robot architecture. "Accountability" is part of this type of systems, as the author states: "The theory of justice argues that there must be a trail back to the responsible parties for such events. While this trail may not be easy to follow under the best of circumstances, we need to ensure that accountability is built into the ethical architecture of an autonomous system to support such needs". The architecture proposed by Arkin [8] is not known to have been implemented.

This type of architectures are based on internal models of the environment where the consequences of the decisions taken by the robot can be evaluated. In [5], Stage simulator is used to evaluate the actions of a "minimally ethical robot" while operating in a simplified environment. In this way, the decisions of the robot could be explained according to the simulations made. [4] formalizes this technique for developing verifiable ethical components.

These type of simulations could be used to "explain" how the robot chose an action, but from our point of view, they can be included in the traditional planers used in deliberative architectures for autonomous systems, as we will describe in next section.

The rest of this paper is organized as follows. Section 2 describes our proposal, first at the cognitive level and then at the development framework level. In order to illustrate the proposed ideas we use the HiMoP [13] architecture and *de facto* standard for the development of software for mobile robots (ROS: Robotic Operating System). Section 3 describes two alternatives for the implementing an accountability module. Section 4 discusses the use of the both proposals in a basic scenario. Finally, Sect. 5 summarizes the conclusions and the further work.

2   An Integrated Accountability Module

The accountability process is intended to be fair, transparent, explainable and accountable. Regarding fairness, accountability has to be free from bias, dishonesty, or injustice. It is also important for an accountability process to be transparent for the robot operation. Finally, it has to be explainable to ensure that every action the robot takes may be audited. In order to fulfill these requirements, we think that modules or components that would implement the accountability mechanisms have to be integrated in the development frameworks used. On the other hand, we also think that this module has to be considered in the cognitive part of the control architecture.

The design of the accountability mechanisms has to take into account the resources (computational, storage, communications, etc.) that would be devoted to these mechanisms. The customization also should let designers define the balance between performance and accountability. This decisions probably will be based on the mission of the robot, the requirements would not be the same for a security robot (capable of engaging with potential threats), that for a small marketing robot (a smart mobile screen).

Below we present these two dimensions in more detail. Regarding accountability at cognitive level we use the HiMoP architecture [13] to illustrate our proposal. Relating to Accountability at development framework level, we present our ideas over ROS framework [15].

2.1   Accountability at Cognitive Level

The main mission of cognitive architectures is the selection of actions or behaviors, see [17]. Many different cognitive architectures have been proposed and implemented for mobile robots, from the ones based on general planners as Soar [19] or ADAPT [10], to the ones closer to neural-inspired systems as the model presented at [18] or ethologically-inspired as Motor Schema [9].

One popular taxonomy of the robot architecture is the one based on the level of abstraction managed, which classifies them into Deliberative, Reactive or Hybrid. We think that all of them should include accountability capabilities. That is, data about how robot behavior was chosen need to be recorded in order to facilitate the behavior explanation, or to implement services for forensic analysis.

For instance, the HiMoP architecture proposes a hybrid cognitive architecture based on three components: a Hierarchy of needs to define robot drives; a set of Motivational variables connected to robot needs; and a Pool of Finite-State Machines (FSMs) to run robot behaviors.

Figure 1 summarizes HiMoP main components. The Deliberative Subsystem determines how robot behaviors are organized to accomplish the long-term goals of the robot. The Reactive Subsystem provides a quick robot answer to respond unexpected environment changes. Finally, the pool of FSMs presents each robot behavior as a FSM defined by states and transition arcs.

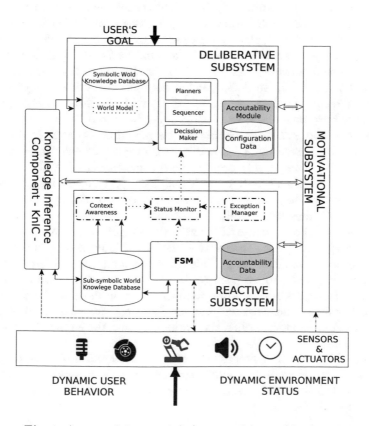

Fig. 1. Accountability module in a cognitive architecture.

Elements that make up the accountability module are coloured in yellow at Fig. 1. Parametrization of the accountability module is included in the Deliberative Subsystem. This configuration indicates which data should be registered to facilitate the behavior explanation. The accountability data themselves are part of the Reactive Subsystem, where low level data is processed.

In summary, first, the accountability module must record information regarding how abstract (high level) decisions were taken, which is made at the deliberative level. These decisions are informed by raw data managed at the reactive level, so secondly, the accountability module must record raw data (low level) during robot operation, working as a *Black-Box-like* system, according to the parametrization made.

2.2   Accountability at Development Framework Level

The conceptual difference between decision making and logging was explained in previous section. But, how we argued in the introduction, the accountability module should be integrated in the *framework* used to implemented the cognitive architecture.

Using the previous example, HiMoP architecture has been implemented using the ROS framework. ROS implements a message-passing distributed system where computation takes place in processes, named *Nodes*, which can receive and send *Messages* into information buffers called *Topics*. We think that ROS framework should provide tools for an easy implementation of accountability.

Thus, it is necessary to decide *which* accounting data are required, *who* is in charge of recording them, and *when* to do it. In addition, it is important to know where, how long, and how the data may be stored.

Regarding who will be in charge of the recording process, in order to not modify the ROS framework (if possible), we propose to use a middleware. We will refer to it as Black Box. In order to do so, this module should run every time a publisher node sends a Message to a Topic. Ideally, the recording process should be transparent to the publisher node, so we need a mechanism to encapsulate Black Box in the messages publication process.

Black box will be in charge of recording messages every time publisher node sends a message to a topic. To ensure scalability it is necessary to delimit which data are recorded and how long. This meta-information is different depending on the publisher node, so we need to decide it for everyone. We require a configuration data repository to store meta-information data so the Black Box to access it. Configuration data will include for each publisher node at least the following data: the message type, the periodicity and the circumstances required to record them.

Once decided which messages are going to be stored, we have to decide the moment to do it. Every time a publisher node sends a message to a topic, Black Box will decide if it saves it according to the meta-information gathered for each publisher node.

3   Designing Accountability in ROS

ROS architecture is usually composed by one ROS Master and some clients. ROS Master is the key element in the ROS system. It runs as a nameservice and manages registration information about all topics used by nodes. The Master is updated in real time by the running nodes, which provide information about topics they publish/subscribe and the type of message used by each topic.

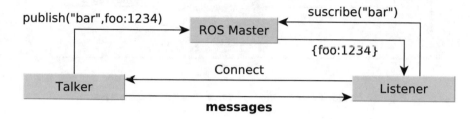

publish("bar",foo:1234)   suscribe("bar")

ROS Master

{foo:1234}

Connect

Talker

Listener

messages

Fig. 2. Conceptual model of ROS topics.

Figure 2 describes the conceptual model of ROS, showing the Master with two nodes, Talker and Listener. Node Talker publishes messages of type *foo* in the topic *bar*, while node Listener is subscribed to topic *bar* to receive its messages. Unfortunately, no security, nor accountability were considered in the design of the distributed communication mechanism of ROS. There are some recent works to improve security in ROS [11], but they do not implement accountability up to our knowledge.

[12] proposes and tests a modification to the ROS organization for improving the cybersecurity of robotic systems. This modification included a specialized node to check the safety and cyber-security of the system. This system was configurable using semantic rules. The accountability module that we propose is somehow similar to the cybersecurity solution presented in [12]. We intend to use a superimposed layer to manage accountability by monitoring the nodes pretending to publish messages and forcing them to save some accountability data.

Figure 3 shows the solution proposed. In this regard, main issues to deal with has to do on which accounting data are required, who is in charge of recording them, when to do it, how long, and how the data may be recorded. Different aspects are described below:

Which Messages on topics included at configuration data repository.
Who Accountability module.
When Every time a publisher nodes sends a message.
How long As much time as indicated at configuration data repository.
How We propose to use the *Rosbag* files. These files are a well-known and commonly used mechanism to store data for the ROS community.

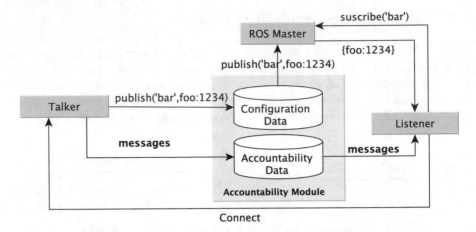

Fig. 3. Accountability module in the robot operating system.

Performance and scalability need to be considered at this point. Number of messages exchanged by the ROS platform can be large, and data processed by the different modules huge; consider for instance the point clouds capture by a RGB-D sensor during robot operation.

Regarding to the above, it is important to note that every publisher node in ROS sends specific messages at a given rate. It is important to decide which messages have to be preserved, and for how long, in order to limit the amount of stored data and thus to guarantee scalability of the system, but also ensuring there is enough information to be capable of describing robot behavior offline. For instance, having information about decisions taken by the robot or destination locations for navigation looks pretty much important than having all data about the odometry at every time. So we may store all decision-making or navigation data for a long time and odometry data just in certain moments when something unexpected occurs.

As mentioned at Sect. 2.2 a repository is required to store meta-information about every node running. For each publisher node, we need to decide the messages we need to store, how long and under which circumstances store them. Decision has to be taken when a new publisher node registers on ROS Master. This decision has to be notified to the process in charge of data recording. An easy approach is to use an YAML file for the configuration data repository. YAML files are commonly used by the ROS community for configuration stuff. This file should be written every time a node register or update on ROS master and read by the process in charge of data recording.

We propose two possible implementations for the above solution, both trying to minimize changes affecting the ROS core system:

1. RAW logging by rules using the ROSRV framework.
2. Accountability by design. We propose to change the Publisher Class included in the ROS framework.

3.1 ROSRV-Based Implementation

For the first alternative, we propose using ROSRV [14]. ROSRV is a runtime verification framework integrated seamlessly with ROS. It aims to monitor safety properties and enforce security policies. Its core is a monitoring infrastructure that intercepts, observes and optionally modifies messages passing through the ROS system. ROSRV checks system's runtime behavior against user-defined properties, triggering desired actions when verifying these properties. It also controls system state and execution of commands by enforcing an access policy to address security concerns. ROSRV allows to ensure safety properties based on event sequences.

ROSRV is designed to address the safety and security issues in ROS-based robot applications. Figure 4 shows its architectural overview. The main difference from ROS is the RVMaster node, which acts as both a secure layer protecting ROS Master and as a functional layer for protecting the safety of the application: all node requests to ROS Master can be intercepted by RVMaster and all messages can be monitored, and thus the desirable safety and security policies enforced.

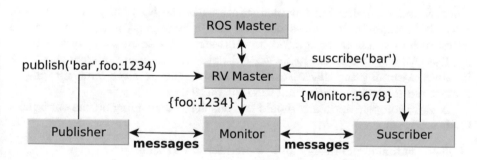

Fig. 4. ROSRV architecture.

Event handlers, called monitors, are defined with rules using a C++ based syntax. For automatic monitor generation out of formal specifications, ROSRV depends on ROSMOP framework which generates code for monitors from the defined rules.

The accountability module, more specifically the Black Box, might be implemented as a ROSRV monitor applying every registered node.

Some changes are required at RVMaster node. When a node tries to register, meta-information about accounting data has to be included at the configuration data repository in order to Black Box is able to record data from the new node.

It is important to note that this change does not affect the ROS core system since ROSRV is a superimposed layer over the ROS framework.

3.2   ROS Modification

For the second proposal, changes in the ROS core system need to be included. Specifically, changes on the Publisher class in the roscpp API are required.

Roscpp API provides a client library that enables C++ programmers to quickly interface with ROS Topics, Services, and Parameters, see [16]. Roscpp API is the most widely used ROS client library and is designed to be the high-performance library for ROS. Publisher class is a virtual class every publisher node must heritage from. It has a Publish method in charge of serialize and send messages. This method has to be rewritten to include some sentences to store messages in the Black Box in addition to send them to their destination topic. To know which messages are required to store, configuration data repository should be read at this point.

To keep configuration data updated, the node registration process need to be also changed. Including meta-information in the configuration data repository for each new node intended to publish messages.

3.3   Scalability

Another issue relating both implementations is the scalability of Black Box. Record messages in time does not look like a very good idea. A method to manage data needs to be applied. Several strategies to manage data may be applied. An easy approach removing old data can be easily implemented. For instance, assuming the Black Box to use Rosbag files to store data, filtering to remove old messages using a timestamp is quite easy.

A more complex approach should be considered depending on the data gathered. For instance:

- Decision making data should be stored for a long time.
- Odometry or sensor data should be stored only under certain circumstances when something unexpected occurs.
- Regarding navigation, probably just destination goals are required to explain robot behaviour rather than full details about the path followed.

4   Discussion

The above proposal to build an accountability module for the ROS framework consists on recording ROS messages published by nodes, according to given meta-information. Messages are a key concept in the ROS framework which allows nodes consuming them to perform the task they have been designed for. For instance, having got data such as obstacles recognized by a navigation node, and the navigation goals generated by a ROSPlan node, it is possible to explain why a mobile service robot behaved the way it did.

Two different implementation proposals were described in the previous section, each one having advantages and disadvantages. The first one presented a ROSRV-based solution. Main advantages of this solution are the following:

- ROSRV allows to implement the black box as a superimposed layer to the ROS framework. In addition to not requiring changes in the ROS core system, this architecture allows to use any ROS package without any modification at all.
- To enforce Security Policies, ROSRV defines access control based on a user-provided specification of access policies as input configuration. On receiving any XMLRPC request, RVMaster decides whether the request is allowed to go to the ROS Master according the specification. The policies are currently categorized into four different sections: Nodes, Suscribers, Publishers, and Commands. Under each section, the access policy is written as a key followed by an assignment symbol and a list of values.

We also need to point out the disadvantages of using a ROSRV-based solution:

- The main limitation of the current implementation is the reliance on IP addresses in particular and on network routing in general to guarantee security. Naively trusting IP addresses does not protect against attackers who can run processes on the same hosts as trusted nodes, or spoof packets on network segments carrying unencrypted traffic.
- Another important drawback regards scalability. Currently ROSRV is centralized. All the monitor nodes live in the same multithreaded process, and all communication in the system is monitored. It supposes a limitation for the message-passing distributed system that ROS implements.

Second proposal describes a brief modification in the ROS core system to implement the accountability module. Main advantages of using this solution are the following:

- The main advantage is that changes are totally transparent for ROS community. At the most, ROS package developers have to provide only a recommendation of the data to be stored in the Black Box for new publisher nodes. Regarding existing ROS packages, they may be used without any changes.
- Unlike ROSRV framework, a centralized solution, this proposal facilitate scalability since it is integrated in the message-passing distributed system that ROS implements.

Regarding the disadvantages of the ROS modification, we must point the following:

- Changes in the ROS framework are required. This is an important drawback because it may need a parallel develop branch for the framework, unless changes are adopted on the official develop branch.
- Unlike ROSRV framework, this proposal does not consider any safety and security issue. If required, they need to be explicitly implemented and evaluated, see [11]. A new version of the ROS framework [20] is being developed to fix security shortcomings, but it has not been officially released.

In addition to the accountability module implementation, we also need to consider some semantic issues. Messages can be enough for explaining at the cognitive level what have happened. For instance, detecting a person in the calculated route may suppose the robot to take a longer path. Even the data absence may explain some events. For instance, a robot may hit a person if he does not recognize her in his way.

A "Black Box" as the one described in this work, may be used as a key clue for a forensic analysis trying to explain why a service robot acted in a certain way. The reliability of the behavior explanation will be related to the amount of data stored in the black box. So the mechanism to decide which messages, under what circumstances, and for how long store them becomes critical. A compromise solution between scalability and the amount of stored data is required.

The accountability solution proposed is focused on messages because they are ROS' main information unit. But there are other levels to be considered, specially those related with the robot decision making. Robot decision making does not necessarily requires sending a message to a specific topic.

Some tasks that are not easily implemented using the "subscription" abstraction. For instance a planner (i.e. ROSPLAN) sends requests to nodes to perform some tasks, and also receive replies to these requests. This is implemented via ROS *services*. In some cases, however, if the service takes a long time to execute, the user might want the ability to cancel the request during execution or get periodic feedback about how the request is progressing. These actions should also be recorded by the accountability module.

ROS also supports servers in charge of accomplishing long-running goals. These servers can be preempted, for instance moving the robot to a distant location, making a 3D-LIDAR scan and returning the resulting point cloud, detecting the handle of a door, etc. ROS `actionlib` package provides tools to create these type of servers. It also provides a client interface in order to send requests to the server. In spite of those communications, usually ending with a node sending a message, it might be difficult to explain a specific behavior only with the final message. `actionlib` package should be improved to provide additional information to the accountability module.

5   Conclusions

We think that as service robotics get higher Technology Readiness Levels (TRLs), software development frameworks will become more standardized, and accountability components as described in this paper will be required. We have reviewed the literature and we have not found tools for implementing accountability for autonomous robots. We believe that an standardized accountability system would facilitate behavior explanation and forensic analysis of safety and security incidents caused or suffered by service robots.

We have justified why we think that accountability has to be implemented as an standard component of the development framework, but that it has also to be included in the cognitive architecture. We have described two alternatives to implement an accountability module for ROS framework. We chose ROS to illustrate our proposal because it is the standard framework for autonomous robots. Implementations described represent a first attempt to implement a standard module for accountability, but they have to be enlarged covering ROS *services* and *actionlibs*.

Further work will be the release of automatic tools that could verify that a particular system has got an accountability module and capable of performing automatic tests in order to certify software controlling autonomous robots. These same tools could also be used in forensic analysis.

Acknowledgments. This work has been partially funded by Junta de Castilla y León grant LE-028P17, Comunidad de Madrid grant RoboCity2030-Fase 3 - S2013/MIT-2748, and by Ministerio de Economía and Competitividad of the Kingdom of Spain under RETOGAR project (TIN2016-76515-R).

References

1. Wachter, S., Mittelstadt, B., Floridi, L.: Transparent, explainable, and accountable AI for robotics. Sic. Rob. **2**(6) (2017). https://doi.org/10.1126/scirobotics.aan6080
2. Delvaux, M.: Report with recommendations to the Commission on Civil Law Rules on Robotics (2015/2103(INL)), January 2017
3. Nevejans, N.: Study: European Civil Law Rules in Robotics. European Parliament PE 571.379 (2017)
4. Dennis, L.A., Fisher, M., Winfield, A.F.T.: Towards verifiably ethical robot behavior. In: AAAI 2015, Artificial Intelligence and Ethics (2015)
5. Winfield, A.F., Blum, C., Liu, W.: Towards and ethical robot: internal models, consequences and ethical action selection. In: Mistry, M., Leonardis, A.,Witkowski, M., Melhuish, C. (eds.) Advances in Autonomous Robotics Systems, volume 8717 of Lecture Notes in Computer Science, pp. 85–96. Springer (2014)
6. Wachter, S., Mittelstadt, B., Floridi, L.: Why a right to explanation of automated decision-making does not exist in the general data protection regulation, 28 December 2016. International Data Privacy Law (2017). https://doi.org/10.2139/ssrn.2903469
7. Arkin, R.C.: Governing lethal behavior: embedding ethics in a hybrid deliberative/reactive robot architecture. Technical report GIT-GVU-07-11. Georgia Tech University (2007)
8. Arkin, R.C.: Governing Lethal Behavior in Autonomous Robots. CRC Press (2009)
9. Arkin, R.C.: Motor schema - baed mobile rsobot navigation. Int. J. Robot. Res. **8**(4), 92–112 (1989)
10. Benjamin, D.P., Lyons, D., Lonsdale, D.: ADAPT: a cognitive architecture for robotics. In: International Conference on Cognitive Modeling (2004)
11. Rodríguez-Lera, F.J., Matellán-Olivera, V., Balsa-Comerón, J., Guerrero-Higueras, A.M., Fernández-Llamas, C.: Message encryption in robot operating system: collateral effects of hardening mobile robots. Frontiers in ICT. Computer and Network Security (2018). https://doi.org/10.3389/fict.2018.00002
12. Balsa-Comerón, J., Guerrero-Higueras, Á.M., Rodríguez-Lera, F.J., Fernández-Llamas, C., Matellán-Olivera, V.: Cybersecurity in autonomous systems: hardening ROS using encrypted communications and semantic rules. In: Iberian Robotics Conference, pp. 67–78. Springer, Cham, November 2017

13. Rodríguez Lera, F.J., Matellán-Olivera, V., Conde, M.A., Martín Rico, F.: HiMoP: a three components architecture to create more human-acceptable assistive robots. Cogn. Process. https://doi.org/10.1007/s10339-017-0850-5
14. Huang, J., Erdogan, C., Zhang, Y., Moore, B., Luo, Q., Sundaresan, A., Rosu, G.: ROSRV: runtime verification for robots. In: International Conference on Runtime Verification, pp. 247–254. Springer, Cham, September 2014
15. Quigley, M., Conley, K., Gerkey, B., Faust, J., Foote, T., Leibs, J., Wheeler, R, Ng, A. Y.: ROS: an open-source robot operating system. In: ICRA Workshop on Open Source Software, vol. 3, No. 3.2, p. 5, May 2009
16. Koubaa, A. (ed.) Robot Operating System (ROS): The Complete Reference, vol. 1. Springer (2016)
17. Hull, C.L., Hovland, C.I., Ross, R.T., Hall, M., Perkins, D.T., Fitch, F.B.: Mathematico-deductive theory of rote learning: a study in scientific methodology (1940)
18. Vinokurov, J., Lebiere, C., Wyatte, D., Herd, S., O'Reilly, R.: Unsurpervised learning in hybrid cognitive architectures. In: Proceedings of the 2012 AAAI Workshop on Neural-Symbolic Integration (2012)
19. Laird, J.E.: The Soar Cognitive Architecture. MIT Press, Cambridge (2012)
20. Thomas, D., Fernandez, E., Woodall, W.: State of ROS 2–demos and the technology behind. Presentation at ROSCon 2015, Hamburg, Germany, October 2015 (2016). https://roscon.ros.org/2015/presentations/state-of-ros2.pdf
21. Halder, R., Proença, J., Macedo, N., Santos, A.: Formal verification of ROS-based robotic applications using timed-automata. In: IEEE/ACM 5th International FME Workshop on Formal Methods in Software Engineering (FormaliSE), Buenos Aires, pp. 44–50 (2017). https://doi.org/10.1109/FormaliSE.2017.9

Robotic Manipulators

Topological Road Mapping
for Autonomous Driving Applications

Esther Murciego[(⊠)], Carlos Gómez Huélamo[(⊠)], Rafael Barea[(⊠)],
Luis Miguel Bergasa[(⊠)], Eduardo Romera[(⊠)], Juan Felipe Arango,
Miguel Tradacete, and Álvaro Sáez

Robesafe Group, Department of Electronics, University of Alcala,
Alcalá de Henares, Spain
{esther.murciego,carlos.gomezh,jfelipe.arango,alvaro.saezc}@edu.uah.es,
{rafael.barea,luism.bergasa,eduardo.romera}@uah.es,
tradacete.miguel@gmail.com
http://www.robesafe.es/personal/bergasa/

Abstract. We present a system to carry out the elaboration of efficient
topological maps for autonomous driving applications by using lanelets
in satellite images provided by an open source tool such as JOSM (Java
Open Street Maps). Our approach is motivated by the need for more
efficient and frequent updates on large-scale maps in self driving applications. The system chains the elaboration of an adjusted topological map
by using JOSM and its subsequent evaluation using both V-REP simulator and real world tests based on a route planning method (Dijkstra) and
trajectory tracking algorithms to implement a ROS based autonomous
electric vehicle able to drive in urban areas.

Keywords: Lanelets · Open Street Maps (OSM)
Java OSM (JOSM) · Dijkstra · Autonomous driving

1 Introduction

Over the past few years, a topic that focuses the attention of automotive companies is the development of autonomous cars. Statistic show that over 70% of
the population in the European Union (EU) is living in urban areas. According
to the World Health Organization, nearly one third of the world population will
live in cities by 2030. Aware of this problem, the Transport White Paper published by the European Commission in 2011 indicated that new forms of mobility
should be proposed to provide sustainable solutions for people and goods safely.
Regarding safety, it sets the ambitious goal of halving the overall number of road
deaths in the EU between 2010–2020. However, the goal will not be easy, only
in 2014 more than 25,700 people died on the roads in the EU (18% reduction).

Autonomous driving is considered as one of the solutions to the before mentioned problems and one of the great challenges of the automotive industry
today. The existence of reliable and economically affordable autonomous vehicles will create a huge impact on society affecting environmental, demographic,

R. Fuentetaja Pizán et al. (Eds.): WAF 2018, AISC 855, pp. 257–270, 2019.
https://doi.org/10.1007/978-3-319-99885-5_18

social and economic aspects. In particular, it is estimated to cause a reduction in road deaths, improved traffic flow, reduced fuel consumption and harmful emissions associated, as well as an improvement in the overall driver comfort and mobility in groups with impaired faculties, like the elderly or disabled people. Autonomous driving has attracted much attention recently by the research groups and industry, due to the billboards of various companies on expectations of market entry. However, his predictions seem to be very optimistic. A more scientific organization, such as IEEE, has recently predicted that by 2040 the majority of vehicles traveling on highways will be autonomous. Driving in urban environments will take longer, due to its complexity and uncertainty.

The autonomous car must be able to navigate without making mistakes, consequently it has to understand the environment. For this proposal, apart from using the most advanced techniques of sensory perception that recognizes the environment in a robust way, it is pretty useful to use a map with topological information of places the car is supposed to be driven.

In this work we address the problem of efficient elaboration and maintenance of topological maps for autonomous driving applications, using an open source map format, easy to integrate and merge with other maps such as OpenStreetMaps [1]. We model relevant parts of the conductive zones with lanelet elements, which are atomic, interconnected and manageable road segments, geometrically represented by a left and right link. Forming a network of nodes with the topological information of the terrain.

The lanelet will not only establish the road and its limits to follow, it will have integrated, the road behaviour that must follow the car too.

The topological mapping approach based on lanelets has been validated in simulation cases using our navigation architecture made in ROS [2] and V-REP simulator [3,4] and our open-source electric vehicle for the real world tests.

2   Open Street Maps

Open Street Maps (OSM) is a collaborative project with the purpose of implementing free and editable maps. These maps are generated by using geographic information caught by mobile GPS, orthoimages and other open-source tools, all this information distributed under Open Database License (ODbL). Registered users can upload their contributions by GPS devices and can introduce or correct vectorial data through the use of edition tools created by OpenStreetMap community. In this work we focus on Java OpenStreetMap editor (JOSM), often used by more experienced OSM contributors when a large dataset must be introduced.

2.1   Java Open Street Maps

JOSM is an editing tool with a similar interface to other traditional geographical information systems (GIS) packages. This application allows the user to edit, tag or import OSM data offline, not being required online connection. For that reason, people all over the world can edit the same zone at the same time through the use of JOSM and then uploading this new data to Open Street Map through the OSM application programming interface (API), being JOSM responsible to merge changes if several people upload different data of the same area. On the other hand, additional services such as linking OSM features, images, audio notes or supporting data conflict resolutions are provided. Moreover, this tool can be improved by using plug-ins, such as RoadSigns and Measurements plug-ins which were set in order to locate different traffic signs related to regulatory elements appropriately.

Main mapping elements in JOSM are:

- **Node:** Point that refers to a given geographical position.
- **Way:** Ordered node list that represents a polyline or polygon (if the polyline starts and ends in the same point).
- **Relation:** Group of nodes, paths or other relations with common properties.
- **Tag:** Fields in which a relation is segmented. Tag sintaxis is divided into a key, a feature and its value. Example: speedlimit $= 40$ defines that this section has 40 km/h as speed limit.

2.2   Lanelets

When creating an efficient map, both in geometric and topological aspects, an accurate description of its elements as well as their relations among them must be supported. Different approaches so as to create this efficient map are:

- **Geometric approach:** Geometric map of the environment.
- **Topological approach:** Abstract description of the map, including the different relationships among its members.
- **Geometric and topological approach:** Merging above approaches.

For that reason, we choose a relatively novel concept called lanelets [5]. This method was proposed originally by Die Freie Universität Berlin with the purpose of drawing an efficient modifying, creating and maintenance map according to a mapping technique able to merge both geometric and topological approach.

A lanelet is defined as an atomic lane segment featured by its left and right bound. These bounds are polylines, and therefore allow for arbitrary precise approximation of lane geometries. The role of a bound, both left and right, specifies the driving direction.

In order to enable routing through a map, a graph made up by adjacent lanelets must be generated. This one can be carried out in two ways: Let the polyline be located in the center of the road or create two polylines that delimit a lane in which the car must be kept, circumscribing the driving zone.

However, if the first option is chosen, a problem arises: The car does not take into account the lane limits, crucial error in autonomous driving applications. For that reason, throughout this project the external campus of the University of Alcala was generated by using JOSM tool with two polylines per each lanelet, each one representing both left and right bound, as illustrated in Fig. 1.

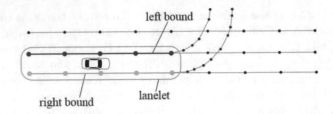

Fig. 1. Map composition by using lanelets. Notice that each lanelet is made up by a left and right bound.

2.3   Mapping Techniques

JOSM offers different choices so as to map our Campus, such as Bing aerial image, Mapbox satellite image, Spain Cadastre or Spain satellite image, among others. For our purposes, since the autonomous car requires an accurately precision, the Spain satellite image, also called PNOA, was used due to its high resolution.

The Plan Nacional de Ortofotografía Aérea (PNOA) has as aim getting aerial orthoimages with resolution from 25 to 50 cm and high precision digital models of elevation of the whole Spanish territory, with an update period spanning from two to three years, depending on the zones. It is a cooperative project cofinanced by General Administration of the State and Autonomous Communities. Figure 2 shows the JOSM interface and our manually two dimensions mapped external campus of the University of Alcala based on PNOA satellite image. In an easy way both lanelets, regulatory elements and different stations can be added or modified. Figure 2 shows from bottom to top several windows were relationships are shown, and the same one to layers and members. Notice that though the format file is OSM, these data are structured in an XML format which provides great portability and easy access to the different members of the map.

However, it is pretty often that the JOSM tool shows a misaligned aerial image, representing a hazardous matter if we pretend to define clear and accurate lanelets where the autonomous car must be driven. In order to solve this problem, we introduce a contribution based on multi-DGNSS positioning integrated in the car with the purpose of verifying the accurate of this satellite image.

The main module of the localization systems consists of a multi-constellation system (multi-GNSS) with RTK positioning solution. This module is directly integrated in the back of the car, made up by two elements: Differential Hiper Pro GPS + receiver configured as rover and a local base station with the purpose of generating differential corrections for the rover.

The rover can obtain data from both GPS and GLONASS to provide a more robust solution than standard GPS by increasing the number of visible satellites. It is able to provide information at a frequency of 10 Hz through the use of differential corrections in order to improve the required accuracy, since autonomous vehicles demand real-time information of the surroundings.

In addition, we set a local based station based on Choke-Ring Antenna, specifically chosen to deal with multipath, connected to a second Hiper Pro

Fig. 2. JOSM interface. It can be observed the external campus of the University of Alcala manually mapped by using lanelets technique

GPS + receiver that provides these differential corrections. As mentioned, these corrections are generated at 10 Hz, published over Internet by using standard open source software in the autonomous vehicle using a GPRS link via radio.

With these differential corrections working, we mapped those zones which were not updated in the PNOA satellite image or even checked, if required, those zones whose drawn lanelets were not matched to these differential corrections projected into JOSM.

Moreover, several fields useful for our ROS based autonomous vehicle architecture were tagged, both for simulation and real world, being found in all lanelets in order to take into account in an easier way the behaviour that the car must carry out. The lanelet fields are made up by:

– **Tags:**
  - **Layer:** Set of members to which a lanelet belongs within the map.
  - **Speedlimit:** Speed to which a road is limited within the map.
  - **Type:** Kind of each member. In this work it is defined as lanelet, station or regulatory element.
  - **Role:** Tag that models the behaviour of a given lanelet. In this work, the following tags are chosen: merging, split, intersection and roundabout.
– **Members:** Elements that make up a lanelet.
  - **Polylines:** Geometric limits that define both left and right bound of a lanelet.
  - **Other relations:** Regulatory elements or stations.
  - **Node roundabout:** Node required if a lanelet is featured with a role roundabout. It can be found in the center of the roundabout.

Regulatory elements were defined as followed:

– **Tags:**
  - **Type:** Regulatory element.
  - **Maneuver:** Behaviour carried out when this regulatory element appears:

    * Give way (give_way)
    * Pedestrian crossing (pedestrian_crossing)
    * STOP
    * Traffic light
   • **Layer:** Set of members to which the regulatory element belongs within the map.
 – **Members:** Elements that make up a regulatory element.
   • **Stop line:** Topological line that sets the limit at which the car should stop.
   • **Reference (ref):** Node that represents the traffic signal of a given regulatory element. If it is a traffic light, it will have more than one reference point: the reference point of its own traffic light and the other reference points of those traffic lights with which the first one is synchronized.
   • **Activate Point (activate_point):** Point in which the car is warned that a few meters from it, there is going to be a signal reference.
   • **End point (end_point):** Point located a little farther than the stop line indicating the maximum point that can be reached if for some reason the stop line is exceeded. It is only defined in semaphores.
 – **Stations:** Possible stopping points for the car.
   • **Tags:**
    * **Type:** In this case it is equivalent to 'station'.
    * **Name:** Name of the station. There are 27 stations in our map.

3   Route Planning

So as to obtain an accurate route planning, first a self-code algorithm for testing lanelets connectivity must be run. Both connectivity checking and Dijkstra algorithm are based on "liblanelet" library [6], available online.

This connectivity is used to check if the relationships among lanelets are correct. The code goes through the nodes of the lanelets and focuses on those that are located at the beginning and at the end of each lanelet. The code analyzes which is the lanelet to which a given node is the end and which is the lanelet to which the same node is the beginning, returning as output that both lanelets are connected.

By using this method it can be checked that all lanelets have at least two connections. This quickly detects which and where are located wrong lanelets, since, if exist wrong connections, the subsequent routing algorithm can not be executed.

In our map all lanelets are connected except for the Polytechnic School. This building has both an absolute beginning node and an absolute ending node, that is, only are related with a single lanelet.

On the other hand, despite of topological and geometrical features of previous lanelets, routes calculation and navigation trajectories are not developed by using JOSM since there are regions that have not got enough data to be reliable. However, there are some searching algorithms based on graphs, such as Dijkstra algorithm [7]. For our purpose we chose the Dijkstra method or minimum path algorithm for its efficiency and speed in our working environment.

This algorithm determines the shortest path, given a vertex of origin to the rest of vertices. The graph is made up by nodes separated from each other by different ways and each one with its own name. Depending on the lanelet length a weight is assigned to each one, obtaining a weighted graph required as input to be useful for this algorithm, as illustrated in Fig. 3. The longer a lanelet is, the more it will weight. Each stop has its own coordinates and through the use of the Euclidean distance we get the closest lanelets, which are used as an argument to get the shortest route with the function shortest_path, of the libLanelet library.

[Distance, Predecessor Node] (Iterations)

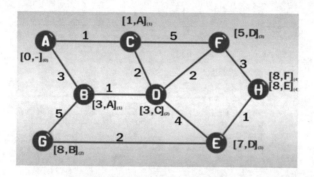

Fig. 3. Dijkstra applied in a graph

4 Simulation Enviroment

Our map was validated by using both RVIZ [8], simulator that belongs to ROS, in order to monitorize the topics and specify the route, and V-REP [3], property of Coppelia and compatible with ROS, simulating the different behaviours cases, such as pedestrian crossing, stop or given way, apart from simulating the vehicle dynamics properly.

V-REP allows importing mesh formats, so an OBJ file is required. Therefore, our map, that is in OSM format, was transformed into OBJ format by using an open source tool called OSM2World.

The visual part of the car seen in simulation has been designed according to our real prototype (Fig. 4), as we show in Fig. 5, imported from a CAD repository adjusting all parameters to this real prototype. It is made up by different car parts as: four wheels with cushioning, propulsion engine, brakes, steering, sensors and a curve base structure based on cuboids.

To obtain a more realistic simulation, the vehicle steering system was simulated in V-REP as a dynamic system, fed up by a servomotor controlled in position. The steering wheel position is internally controlled by V-REP using a PID controller.

In addition, we model different sensors carried by the vehicle, with the purpose of perceiving the environment, such as a LiDAR, a GPS and stereo camera.

Once a map based on lanelets is provided, we use RVIZ simulator, connected to VREP by code, in order to specify the start and goal point of our route, parameters which can not be defined in V-REP. After defining in RVIZ a start and goal point, a series of equidistant points are generated in the centreline of the respective lanelet and the purepursuit algorithm [9] is used to follow this discretized trajectory. Two navigation ways (Fig. 6) can be executed to follow this path:

Fig. 4. Our real prototype equipped with sensors

- **Non-reactive navigation:** Navigation way that only executes the tracking algorithm, without taking into account neither lanelets boundaries nor behaviour cases, such us pedestrian-crossing, give way or stop. It describes a constant radius that causes the vehicle moves to the next target point. The lookahead is adjusted in order to decrease the car speed if the car detect great curvatures, just like a human driver would do.
- **Reactive navigation:** This navigation way takes into account previous laws of Non-reactive navigation but furthermore it is able to face and react both to lanelets boundaries and obstacles, such as pedestrians or vehicles, both in static and dynamic, by using the Curvature-Velocity Method (CVM) [10] and precision tracking algorithms [11], drawing arcs of dynamic radius that describe the possible trajectories to reach the target point avoiding the ahead obstacles.

5 Results

According to these two navigation ways, non-reactive and reactive, we designed a series of routes with the purpose of validating our method, first in simulation and second with a real prototype. Table 1 shows different routes all over the external campus of the University of Alcala with a moderate degree of complexity presenting streets with different speeds, curves and various traffic signals. Each route presents a number of lanelets related to the total distance. The average distance for a lanelet was 116.125 m.

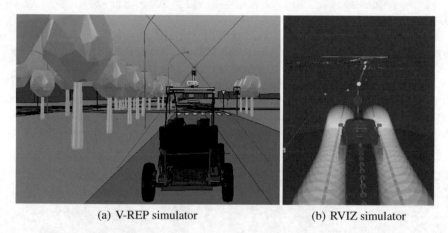

(a) V-REP simulator   (b) RVIZ simulator

Fig. 5. Figure (a) shows our imported car in simulation by using V-REP simulator. Figure (b) shows route planning in R-VIZ simulator

Table 1. Routes used in simulation

Route	StartPoint	EndPoint	Metres	N° lanelets
1	PolytechnicStart	Train Station	1800	20
2	Train Station	PolytechnicEnd	2500	32
3	Biology	University Hospital	1700	8
4	Universitary Residence	CGD Telefónica	2500	25
5	Pharmacy	Polytechnic	1800	18

On the other hand, Table 2 shows the number of roundabouts per each route, its average speed, total time of the route and its respecting navigation way. It can be observed that in spite of the fact that the route 2 and 4 are longer than route 1, because route 1 presents more curves and reactive control is taken into account, it takes more time driving the car throughout the simulation, the same than in real tests. In conclusion, the bigger the amount of curves taking into account reactive control, the more time it will take driving the simulation.

As example, it is explained the route 1, which starts at the Polytechnic school and ends at the Train Station.

The stations and the route of lanelets followed is: *Startstation: Polytechnic-Start*

Endstation: Train Station

Route: [*<Lanelet -51855>*, *<Lanelet -51851>*, *<Lanelet -51837>*, *<Lanelet -51051>*, *<Lanelet -51221>*, *<Lanelet -51087>*, *<Lanelet -51083>*, *<Lanelet -51085>*, *<Lanelet -51217>*, *<Lanelet -51069>*, *<Lanelet -51139>*, *<Lanelet -51149>*, *<Lanelet -51203>*, *<Lanelet -51205>*, *<Lanelet -51451>*, *<Lanelet -51457>*, *<Lanelet -51475>*, *<Lanelet -51561>*, *<Lanelet -51531>*, *<Lanelet -51533>*]

Figure 7 shows the simulation, both in V-REP and RVIZ, under reactive navigation, at the beginning of the route 1. It can be seen the curvature lines in fuchsia color, target points in blue and the path to follow in green.

Figure 8 shows, from above, in one of the roundabouts of the route. The blue target points are observed and the lanelets are detected as obstacles. In Fig. 9, it can seen the route that the vehicle has followed throughout the Campus, using the simulator. Since we don't see the Campus image as background

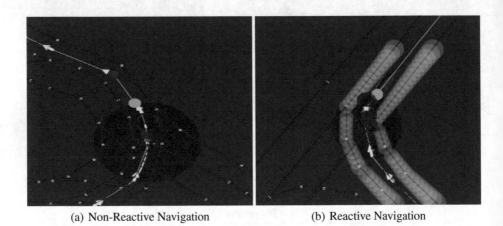

(a) Non-Reactive Navigation   (b) Reactive Navigation

Fig. 6. Figure (a) shows non-reactive navigation and Figure (b) shows reactive navigation

Table 2. Quantitative parameters of routes

Route	Roundabouts	Average speed (km/h)	Time (Seg)	Control
1	4	22.6	491	Reactive
2	5	23.4	266	No reactive
3	4	21.9	290	No reactive
4	7	23.7	284	No reactive
5	5	20.8	208	No reactive

in Fig. 9, we show the followed route most clearly in Fig. 10. As supposed, the shortest route has been chosen. This route has been simulated using the reactive navigation, whose processing time is greater that the non-reactive because non-reactive doesn't have to calculate the obstacles that appear in the route to define its path and its speed. In this case the duration of the route is about 8 min and 11 s, as illustrated in Table 2.

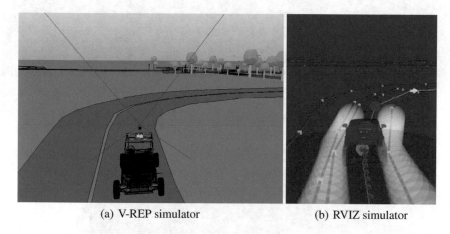

(a) V-REP simulator   (b) RVIZ simulator

Fig. 7. Back view on the route Polytechnic school - Train Station.

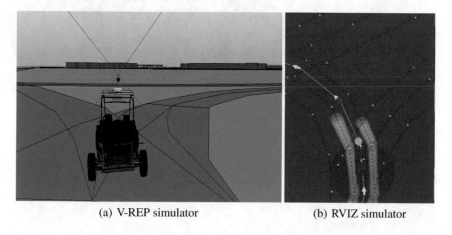

(a) V-REP simulator   (b) RVIZ simulator

Fig. 8. Roundabout situation in the route Polytechnic school - Train Station.

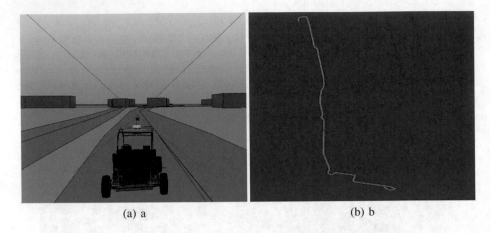

(a) a   (b) b

Fig. 9. Global view of the driven route 1 by using V-REP

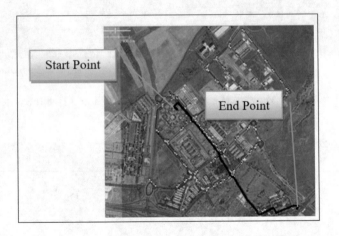

Fig. 10. JOSM view of the route 1: Polytechnic school to Train Station

Finally, Fig. 11 shows different cases both in simulation and with our real prototype following the route Polytechnic school - CGID Telefonica.

Fig. 11. Different situations found in the route 1, both with our real prototype (left column) and in simulation (right column)

6 Conclusions

The implementation of the lanelets map based on the external campus of the University of Alcala has contributed to the final objective of the SmartElderlyCar project [4] which consists of the development of an electric vehicle capable of driving autonomously by our campus. The OSM map doesn't provide enough data to achieve autonomous driving, so it has to be enriched by lanelets that delimit the lane with two polylines thus preventing the car from moving outside the limits established by the road.

In addition, with the method of lanelets it can be included in the same map regulatory elements for the car. The proposal has been validated in simulation using V-REP under ROS by two navigation methods (reactive and non-reactive) and in real tests obtaining successful results.

Acknowledgment. This work has been partially funded by the Spanish MINECO/ FEDER through the SmartElderlyCar project (TRA2015-70501-C2-1-R), the DGT through the SERMON project (SPIP2017-02305), and from the RoboCity2030-III-CM project (Robótica aplicada a la mejora de la calidad de vida de los ciudadanos, fase III; S2013/MIT-2748), funded by Programas de actividades I+D (CAM) and cofunded by EU Structural Funds.

References

1. Haklay, M., Weber, P.: OpenStreetMap: user-generated street maps. IEEE Pervasive Comput. **7**(4), 12–18 (2008)
2. Quigley, M., Conley, K., Gerkey, B., Faust, J., Foote, T., Leibs, J., Wheeler, R., Ng, A.Y.: ROS: an open-source robot operating system. In: ICRA Workshop on Open Source Software, Kobe, Japan, vol. 3, p. 5 (2009)
3. Rohmer, E., Singh, S.P., Freese, M.: V-REP: a versatile and scalable robot simulation framework. In: 2013 IEEE/RSJ International Conference on Intelligent Robots and Systems (IROS), pp. 1321–1326. IEEE (2013)
4. Otero, C., Paz, E., López, J., Barea, R., Romera, E., Molinos, E., Arroyo, R., Bergasa, L.M., López, E.: Simulación de vehículos autónomos usando v-rep bajo ros. Actas de las XXXVIII Jornadas de Automática (2017)
5. Bender, P., Ziegler, J., Stiller, C.: Lanelets: efficient map representation for autonomous driving. In: 2014 IEEE Intelligent Vehicles Symposium Proceedings, pp. 420–425. IEEE (2014)
6. Högdahl, J.: Design specification autonomous trucks (2014)
7. Brandes, U.: A faster algorithm for betweenness centrality. J. Math. Sociol. **25**(2), 163–177 (2001)
8. Gossow, D., Leeper, A., Hershberger, D., Ciocarlie, M.: Interactive markers: 3-D user interfaces for ros applications [ROS topics]. IEEE Robot. Autom. Mag. **18**(4), 14–15 (2011)
9. Coulter, R.C.: Implementation of the pure pursuit path tracking algorithm. Technical report, Carnegie-Mellon UNIV Pittsburgh PA Robotics INST (1992)
10. Simmons, R.: The curvature-velocity method for local obstacle avoidance. In: Proceedings of 1996 IEEE International Conference on Robotics and Automation, vol. 4, pp. 3375–3382. IEEE (1996)
11. Oron, E., Kumar, A., Bar-Shalom, Y.: Precision tracking with segmentation for imaging sensors. IEEE Trans. Aerosp. Electron. Syst. **29**(3), 977–987 (1993)

Design and Implementation of a Low Cost 3D Printed Adaptive Hand

Alberto J. Tudela, Joaquin Ballesteros, and Antonio Bandera[⊠]

Dpto. Tecnología Electrónica, E.T.S.I. Telecomunicación,
Universidad de Málaga, Málaga, Spain
ajbandera@uma.es

Abstract. Service robot are acquiring significant importance on every-day scenarios. They are able to provide relevant services to human beings like household tasks or helping senior people to live independently. Nevertheless, some tasks, such as cooking or grasping objects cannot be provided without equipping the robot with robotic arms. However, on many occasions, the price of the arms exceeds that of the robot itself. Some low-cost solutions have been proposed to solve this problem, but they usually do not adapt the end-effector (gripper) to the grasped object. Based on the design of the InMoov open source humanoid robot, this paper describes the design and implementation of a low-cost arm with an adaptive gripper for humanoid robots. Electronics is deployed in a distributed architecture based on Arduino platforms for low-level control. Obtained results show that the proposed arm is able to successfully solve grasping tasks.

Keywords: Robotic arm · Adaptive grip · Service robots

1 Introduction

Recent developments in robotics show a growing interest in service robots [1,2]. They are able to perform task like cleaning houses [3] and take care of elderly and chronically ill people [4]. It is expected that an increasing number of personal and professional service robots will be designed to provide a continuous interaction with people. For instance, this is the case of the mobile platforms for logistic and transportation in large hospitals [5], or for those designed for guiding people in museums or other large installations [6]. But where this skill is more evident is in robots like the Tiago of PAL Robotics [9] or Pepper of Aldebaran [10]. These robots have been designed for satisfying the requirements (skills and body features) imposed by this interaction. Partially, they mimic the human shape and motions [11], being usually equipped with arms and hands for grasping light objects. Furthermore, these robots are endowed with natural human-robot interaction capabilities [7]. Verbal communication in humans does not come isolated from non-verbal signs. Hence, it is interesting that these robots

© Springer Nature Switzerland AG 2019
R. Fuentetaja Pizán et al. (Eds.): WAF 2018, AISC 855, pp. 271–283, 2019.
https://doi.org/10.1007/978-3-319-99885-5_19

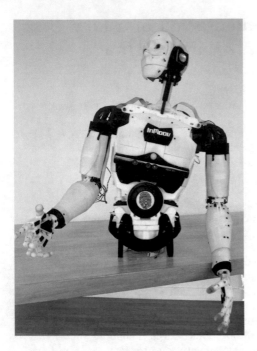

Fig. 1. (Left) The InMoov robot by Gael Langevin (https://inmoov.fr); and (right) a detailed view of the design of the hand (http://www.reppersdelight.spacymen.com).

will be equipped with arms and hands. Together with facial expressions or head gestures, arm and hand gestures are basic in the human-human interaction.

One option for equipped a social robot with arms is to use industrial possibilities. Modern industrial robotics provides numerous examples of manipulator arms, which are able to lift hundreds of kilograms [12], to perform a repetitive task with a precision of tens of micrometers [13] and/or to achieve accelerations up to 15 g [14]. Such programmed and simple tasks are performed in a controlled environment. The increase in manual labor has encourage researchers to develop more advanced robotic arms and grippers. But most of the industrial grippers are not suitable for social robots as they do not adapt to the changing environment in which people move. Advanced grasping of complicated objects is an active research area, being robotic hands continuously redesigned to increase the compliance of grippers in unknown environment. Solutions include the use of vision systems, and flexible and biomimetic grippers. Camera systems are employed to detect the presence of objects and to send their positions to the robotic arm to pick them. Although this system could be mounted on the robot, it will need that the robot can gaze their cameras to the objects and also that it will be able to analyze the scene in an active manner for capturing the needed information (e.g. for grasping the object) [8]. The alternative is the use of external cameras, which is not a practical solution for several reasons [15]. On the other hand, multi-fingered grippers that imitate the behavior of the human hand from a biological perspective [16] is a challenge of design as they are dexterous hands. These robotic hands allow to move the fingers with total precision using a high number of actuators: as many as degrees of freedom have the hand. The cost of the aforementioned systems is typically very expensive.

Reducing the number of actuators letting the joints of the fingers to move freely allows grasping oddly shaped objects. For instance, Kragner et al. [17] built a pair of underactuated hands following this strategy. The solution was valuable due to the adaptability of the gripper to mold around the object it is holding. In the line of low-cost robotic hands [18,19], the open source project InMoov stands out. This project proposes the creation of a humanoid robot in real scale built with a 3D printer (see Fig. 1). Our proposal is based on these two guidelines: underactuation and reduced price. Thus, the whole design is based on the InMoov arm and hand, which has been modified to provide a response to several drawbacks presented in the original design. Our proposal uses low cost forces sensors in a 3D printed robotic hand in conjunction with two Arduino boards, so it can be used with a reasonable cost. It is controlled by five actuators, which pulls a tension cable along the fingers. This design is capable of grasping objects in multiple configurations. The links will flex based on whether the object contacts the distal or proximal links.

The rest of the paper is organized as follows: Sect. 2 describes the proposed system -i.e mechanical design, electronics and software-. Experimental evaluation are described in Sect. 3. Section 4 discusses the obtained results and proposes future work.

2   Design

With the aim of providing a low-cost solution to the problem of adding arms to a social robot, this paper presents a 3D printed human-like robotic hand and arm. The proposal uses pressure sensors that are readily available for purchase. These sensors are mounted in the palm of the hand and covered with a rubber-like flexible 3D printed material, which allows for better gripping ability and also provides a malleable surface that allows the sensors to work optimally. Next sections describes the mechanical (Sect. 2.1) and electronic (Sect. 2.2) design, and the control software (Sect. 2.3).

2.1   Mechanical Design

As aforementioned, the whole mechanical design is based on the InMoov project. However, there are relevant modifications on the forearm and hand. Next Sections describe our proposal. All parts were designed using Autodesk Fusion 360^TM.

Wrist and Gears Set. Figure 2 shows the overall design of the forearm. Using inspiration from InMoov arm, the servomotors that drive the motion of the fingers and the wrist are housed on the forearm. The bed that holds five of these

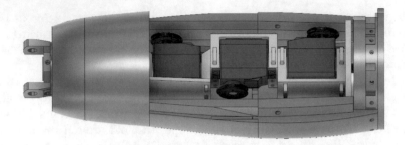

Fig. 2. Overall design of the forearm.

servos has been designed to accommodate standard servos that can be easily replaced. The bicep is designed to hold an Arduino Nano. This microcontroller provides the control actuation. For doing this, it communicates with a board located in the back of the hand, where the information coming from pressure sensors is centralized (see next Subsections).

The motion system of the fingers follows an underactuated tendon-driven scheme. This forces the tendons to pass through the wrist. In the InMoov, with a fixed bed to hold the servomotors, the rotation of the wrist changes the tension on the tendons: by turning the wrist clockwise the tension increases, and, when turning the wrist counterclockwise, it is relaxed. That prevents the use of the hand in various configurations of the object space.

In our proposal, the bed of the servomotors is rotated in conjunction with the hand. The bed was configured to place the servomotors in opposite directions along the forearm (see Fig. 2). At the end of the forearm (the elbow in the human model), a bearing allows the entire system to rotate around the same axis. A set of planetary gears has been placed on the wrist. A ring gear is attached to the forearm and a sun gear is attached to the servo of the wrist. When this servo rotates in clockwise, the hand rotates in counterclockwise and vice versa. The gear ratio of the system is $n_1 : n_2 = 45 : 30$, that is, the hand rotates from $-60°$ to $60°$. Figure 3, shows the planetary gears with the wrist that houses the pipes for the tendons.

Sensors Integration. Pressure sensing is one of the most important human senses. There are pressure sensors of many shapes and sizes. Piezo vibration sensors were considered for their low cost and flexiforce pressure sensors for their size. Low-cost, small size and easy implementation were the most important factors taken into consideration while choosing the sensors.

The pressure sensors chosen were MF01-N-221-A04 and MF01-N-221-A01 from Alpha (Taiwan), and FSR03CE from Ohmite. These type of sensors are force-sensing resistor whose resistance changes when a force, pressure or mechanical stress is applied to them. Therefore, the force is measured by a change in resistance. It was found that these sensors were not as tough as we wanted them to be. In our initial design, it was decided to place the sensors inside the

Fig. 3. Planetary gears that allows the motion of the hand.

Fig. 4. Original design with the sensor inside the finger.

fingers, simulating the fingertips (see Fig. 4). However, after numerous tests, this implementation was discarded when ended up with some broken sensors.

Table 1 provides more details about the employed sensors. As it is illustrated at Fig. 5, three different sizes have been finally included in our design. A relevant feature of these force sensors is that the sensitivity varies with the applied bias voltage. The electronic design is described in Sect. 2.2.

Hand Design. The final choice to endow the hand with sensing capabilities was to integrate the selected sensors into the palm of the hand. To achieve this, the palm was designed to comfortably hold five pressure sensors. As can be seen in Fig. 5, the main part of the hand has been carved out to provide the space where fitting the pressure sensors. However, there is one additional difficulty for using

Table 1. Comparison of the sensors

Model	Force range	Diameter
MF01-N-221-A04	10 g to 1 kg	13 mm
MF01-N-221-A01	30 g to 1 kg	17.5 mm
FSR03CE	20 g to 5 kg	30.5 mm

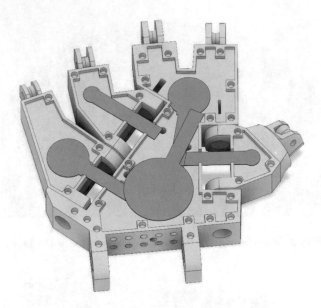

Fig. 5. The palm and the five sensors embedded in it.

these pressure sensors: the small size of its sensing area (13 mm for the smallest and 30.5 mm for the biggest). In order to increase the size of the sensing area, the top of the palm has been designed as a cantilever plate anchored at two ends. This plate is hovered above the sensor, with a small protrusion that concentrates the force applied anywhere on the cantilever onto the center of the force sensor. Figure 6 shows how one cantilever is positioned on one of the regions conforming the palm of the hand. With the addition of the cantilever, the sensing area is increased, allowing to measure the pressure value in practically any place of the palm.

In addition to the cantilever, a rugged and flexible semisphere is added over the cantilever. The semisphere provides a better contact of the cantilevers with the grasped object. The semisphere was also 3D printed using Filaflex filament. Figure 7(a) shows the semispheres on the hand.

Underactuated Tendon-Driven Fingers. The size of the fingers is similar to the one adult human fingers. They are divided as human fingers: distal, middle, and proximal phalanges. To hold the phalanges together, a piece of 3 mm filament is inserted into the joint (Fig. 4 shows the three holes where these filaments are inserted).

As it was aforementioned in Sect. 1, the low-cost needs drove our design towards an underactuated hand with tendon-driven fingers. This approach offers at the same time adequate grasping capability while reducing the overall cost. Adopting this design, the controlled motion of each fingers is addressed using only one actuator per finger. We use five Hobbyking HK15298B servomotors

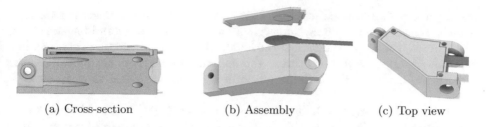

(a) Cross-section   (b) Assembly   (c) Top view

Fig. 6. (a) Cross-section view of the cantilever. Sensor is drawn in grey, the palm in pink and the cantilever in blue. (b) The assembly of this three elements. (c) Final top-view of this part of the palm.

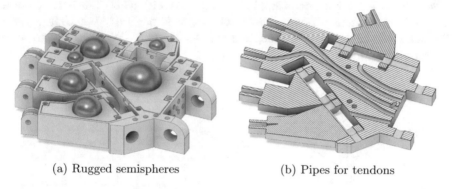

(a) Rugged semispheres   (b) Pipes for tendons

Fig. 7. Different views of the hand showing (a) the flexible semispheres, and (b) the pipes for the tendons

for the fingers and one Towerpro MG996R for the wrist. As aforementioned in Sect. 2.1, these servos are located in the forearm. The standard servo horns were inserted in 3D printed circular horns, eliminating the need for screws.

A stainless steel wire goes from a servo to the tip of each finger, passing through the three phalanges, similarly to a tendon in the human hand does. These wires are wrapped and tied around the servo horns. When the servo rotates, the wire is wrapped further around the horn an pull the tip of the finger in the direction of the hand. Thus, the hand closes as follows. First, the tension of the tendon pulls the distal phalanx and bend it inward. Then, as the tension increases, the middle and proximal phalanges bends, closing the hand. This way, we have 3 degrees of freedom controlled by a single actuator. To open the hand, the servos rotates in the opposite direction, relaxing the tendons. Figure 7(b) shows the pipes carved out on the hand for allowing the tendons go through them.

2.2 Electronic Design

Due to the complexity of bringing the outcomes of the force sensors from the hand to the control board located in the biceps, a distributed architecture has been developed. In this architecture, there is a master board located in the biceps, which is charge of controlling the servomotors that move wrist and fingers, and a slave board located on the back of the hand, which is the responsible of providing the power supply and of capturing the measure values from the sensors. Both boards are based on the Arduino platform and communicated through an I^2C bus.

The sensor board was designed using Autodesk Eagle™ as a shield that can be mounted over the Arduino Nano. It consists of two well-defined parts: a first stage for conditioning the signals coming from the sensors and a second one for filtering and calibrating these signals. The first stage builds a simple current–to–voltage converter as shown in Fig. 8(a). The force sensor resistance (FSR) is powered by $-V_{ref}$. The operational U_1 is adjusted with a 600Ω resistor to obtain the maximum dynamic range. The whole board is powered by a charge pump. The wires providing the I^2C bus, and the $+3.3V$ and GND signals, go inside the wrist as well as the tendons.

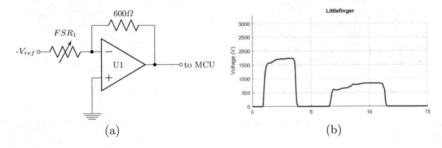

(a)   (b)

Fig. 8. (a) Circuit for conditioning the sensor output signal, and (b) voltage output when the littlefinger sensor is pressed. The first curve is associated to a pression right above the sensor, and the second one when is pressed far away.

The graph in Fig. 8(b) was generated by applying pressure on the littlefinger carpal at two locations: right above the sensor and far off the sensor. As can be seen, the cantilever responds to pressure applied at even in the corners of the carpals. Also shows that the far side of the cantilever deflect less and, therefore, there is a lower voltage response.

2.3   Software Design

This Section briefly describes the software developed for controlling the motion of the adaptive hand for holding an object. The key idea is to open and close fingers, displacing an estimate of the position of the grasped object towards the center of mass of the hand. Using the previously mentioned CAD tool, the center of mass of the hand was calculated. Using the coordinates of this center of mass as reference point, we obtain the coordinates of the sensors. The position of the grasped object is determined by the pressure centroid, C_p, calculated as:

$$C_p = \frac{1}{F} \sum_{k=1}^{5} F_k \cdot (x_k, y_k)$$ (1)

being F the mean of forces, F_k the force on k sensor and (x_k, y_k) its position.
Figure 9 shows the position of a grasped object (red mark) from the center of
mass (black mark). In this case, the object is near the littlefinger, pressing its
sensor.

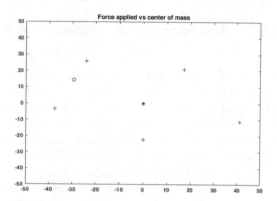

Fig. 9. Position of the object (red circle) from the center of mass of the hand (black
star). Sensor positions are marked with a cross blue.

The control loop goes as follows. For each finger, we move it a certain amount
of angle $\Delta \alpha$[1] and then we obtain the new pressure centroid. This value allows to
calculate the position error of the grasped object, as the difference between the
pressure centroid and the center of mass. If the error has increased, we return
the finger to its original position. The process runs quickly and converges to a
position where the pressure centroid is very close to the center of mass in the
hand. Because the sensors has a low accuracy and low consistency, we set a circle
of 6 mm of radius around the center of mass of the hand. If the pressure centroid
falls inside this circle, it is considered that the object has been correctly grasped
and the servomotors of the fingers hold its positions.

3   Experimental Evaluation

The hand has been qualitatively tested by grasping a large variety of objects,
with different sizes and shapes. Figure 10 shows two examples. The hand is capable of grasping small to medium sized objects. The servos supplied excellent
torque to hold objects up to 2 kg.

---

[1] It has been heuristically set to 5°.

(a) Hand grasping a tool   (b) Hand grasping a multimeter

Fig. 10. Hand grasping objects of different shapes.

Evaluating the performance of multi-fingered, underactuated hands is a challenging task. For fully actuated hands, where the contact points on the grasped object and the corresponding pose of the whole hand can be estimated, form and force closures can be employed for achieving this evaluation. However, the application of these measures for the case of underactuated hands is not straightforward due to the fact that this kind of grasping will adapt the hand shape to the object, being irrelevant to find specific contact points or precomputing the hand configuration. The effective capacity to apply contact forces can be evaluated using two measures: the Potential Contact Robustness (PCR), and the Potential Grasp Robustness (PGR). Pozzi et al. [21] proposed the use of these measures for evaluating the grasping quality for underactuated hands. However, in their proposal, the position of the object in the hand is not considered, but only the applied forces and the number of contact points. Briefly, the PGR establishes that an increasing force applied over a major number of points provides a better grasping. In our case, the presence of pressure sensors allows a direct estimation of the force measures. The inherent uncertainty in the position of the contacts that appear in a real scenario is avoided as our design is able to provide the pressure value in a given region (a practical implementation of the Independent Contact Region (ICR) [22]). Additionally, our proposal also considers how the object is hold in the hand. Moving the center of pressure to the center of the hand is also an index of stability.

The proposed scheme for correctly grasping an object has been assessed with the basic control algorithm of closing all fingers at once. Numerous tests were carried out with various objects, such as a multimeter, a bottle, a cutter, a file and a heat roll. Two batch of tests were performed. In the first batch, the grip of the objects mentioned above was checked with the basic control algorithm. In the second one, the grip was tested out using our control algorithm, which takes into account the sensors placed in the hand. In both cases the objects were always placed in the same hand and in the same position.

The main advantage of the proposed scheme is that it reduces the distance between the pressure center to the center of the hand. Furthermore, it also increases the applied forces. Table 2 summarizes the results after 50 grasping tasks. With the proposed control scheme, the center of pressure is closer to the

Table 2. Basic control versus proposed control algorithm.

Control	Basic		Proposed	
	Mean	St. Dev.	Mean	St. Dev.
Timing of grip (s)	0.36	0.00	5.19	0.02
Average forces (mV)	342.29	77.56	465.15	125.83
Pressure centroid (mm)	31.47	3.26	12.30	4.25

center of the hand, obtaining a mean of the error at the center of pressure of 12.30 mm with respect to 31.47 mm of the basic control; with a slightly higher standard deviation 4.25 versus 3.26. In addition, with the proposed control we get more sensors activated, as shown in the increase of the mean of the average forces and its standard deviation. In contrast, the stabilization time of the entire system has increased considerably (from 0.36 s to 5.19 s).

4   Conclusions and Future Work

This paper proposes the design of arms and hands for a humanoid robot. They have been developed considering the major guidelines of including all electronics, and being lightweight, and low-cost. The physical design is based on the open source 3D printed robot InMoov. However, significant modifications have been addressed on the forearm and hand. Electronics have been implemented for endowed the arm with a sensor board in the hand and a master, control board in the forearm. Thanks to the 3D printing technology, it has been possible to bring very quickly the initial design, and its modifications, from the proposal to the real prototype. Different printing materials have been employed. The Filaflex allows to design flexible elements that, in our prototype, have allowed to cover the palm of the hand with an interesting 'skin', able to provide a pressure measure from practically all the palm of the hand. This allows to avoid the traditional premise that each contact applies a unitary normal force [21], providing a realistic quality measure of the magnitude of the force. Next steps in this analysis will imply to redesign the PGR equation according to our possibilities for measuring. This index will provide a good index for evaluating and comparing different grasping strategies.

Future work focuses on improving the implementation of the grasping strategy, as it is providing good results but it takes an excessive time. The idea is to learn policies for moving the center of the object to the center of mass in a faster way. Reinforcement learning can be used for providing a solution, using as input the pressure maps (Fig. 9). It is also planned to link this control board to a software agent, which allows to incorporate the arm and hand to the CORTEX architecture [20]. The initial aim is to provide our robots with arms and hands for augmenting their verbal interaction channel, and also for addressing simple tasks. Next step will imply to link vision and manipulation skills through an internal representation of the scene stored in an inner world.

References

1. Anandan, T.: Service Robots on the World Stage, Robotics Online. https://www. robotics.org/content-detail.cfm/Industrial-Robotics-Industry-Insights/Service-Robots-on-the-World-Stage/content_id/7061. Accessed 10 July 2018
2. Schraft, R.D., Schmierer, G.: Service Robots. AK Peters/CRC Press (2000)
3. Forlizzi, J., DiSalvo, C.: Service robots in the domestic environment: a study of the roomba vacuum in the home. In: Proceedings of the 1st ACM SIGCHI/SIGART Conference on Human-Robot Interaction, pp. 258–265 (2006)
4. Nicholas, R., et al.: Towards personal service robots for the elderly. In: Workshop on Interactive Robots and Entertainment (WIRE 2000), vol. 25 (2000)
5. Acosta Calderon, C.A., Mohan, E.R., Ng, B.S.: Development of a hospital mobile platform for logistics tasks. Dig. Commun. Netw. **1**(2), 102111 (2015)
6. Thrun, S., Bennewitz, M., Burgard, W., Cremers, A.B., Dellaert, F., Fox, D., Schulz, D. (n.d.).: MINERVA: a second-generation museum tour-guide robot. In: Proceedings 1999 IEEE International Conference on Robotics and Automation (Cat. No. 99CH36288C), 3 19992005 (1999)
7. Mavridis, N.: A review of verbal and non-verbal humanrobot interactive communication. Robot. Auton. Syst. **63**(1), 22–35 (2015)
8. Holz, D., Nieuwenhuisen, M., Droeschel, D., Stückler, J., Berner, A., Li, J., Klein, R., Behnke, S.: Active recognition and manipulation for mobile robot bin picking, In: Gearing Up and Accelerating Cross-Fertilization between Academic and Industrial Robotics Research in Europe - Technology Transfer Experiments from the ECHORD Project. Springer Tracts in Advanced Robotics (STAR), vol. 94, pp. 133–153 (2014)
9. Pages, J., Marchionni, L., Ferro, F. (n.d.).: TIAGo: the modular robot that adapts to different research needs (2016)
10. Tanaka, F., Isshiki, K., Takahashi, F., Uekusa, M., Sei, R., Hayashi, K.: Pepper learns together with children: development of an educational application. In: 2015 IEEE-RAS 15th International Conference on Humanoid Robots (Humanoids), pp. 270–275 (2015)
11. Cheng, H., Ji, G.: Design and implementation of a low cost 3D printed humanoid robotic platform. In: The 6th Annual IEEE International Conference on Cyber Technology in Automation, Control and Intelligent Systems (2016)
12. Siciliano, B., Khatib, O.: Springer Handbook of Robotics. Springer, Heidelberg (2008)
13. Monkman, G.J., Hesse, S., Steinmann, R., Schunk, H.: Robot Grippers. Wiley, Weinheim (2007)
14. Krüger, J., Lien, T.K., Verl, A.: Cooperation of human and machines in assembly lines. CIRP Ann. Manuf. Technol. **58**, 628–646 (2009)
15. Wang, Y., Zhang, G.-L., Lang, H., Zuo, B., de Silva, C.W.: A modified image-based visual servo controller with hybrid camera configuration for robust robotic grasping, pp. 1398–1407 (2014)
16. Xu, Z., Todorov, E.: Design of a highly biomimetic anthropomorphic robotic hand towards artificial limb regeneration, pp. 3485–3492 (2016)
17. Kragten, G.A., Baril, M., Gosselin, C., Herder, J.L.: Stable precision grasps by underactuated grippers. IEEE Trans. Robot. **27**, 1056–1066 (2011)
18. Omarkulov, N., Telegenov, K., Zeinullin, M., Begalinova, A., Shintemirov, A.: Design and analysis of an underactuated anthropomorphic finger for upper limb prosthetics. In: 2015 37th Annual International Conference of the IEEE Engineering in Medicine and Biology Society (EMBC), Milan, Italy (2015)

19. The Open Hand Project. http://www.openhandproject.org/. Accessed 15 July 2018
20. Bandera, A., Bandera, J.P., Bustos, P., García-Varea, I., Manso, L., Martínez-Gómez, J.: CORTEX: a new Cognitive Architecture for Social Robots, Viena, Austria, EUCog (2016). 10.13140
21. Pozzi, M., Sundaram, A.M., Malvezzi, M., Prattichizzo, D., Roa, M.A.: Grasp quality evaluation in underactuated robotic hands. In: 2016 IEEE/RSJ International Conference on Intelligent Robots and Systems (IROS), Daejeon, pp. 1946–1953 (2016)
22. Roa, M.A., Suarez, R.: Computation of independent contact regions for grasping 3-D objects. IEEE Trans. Robot. **25**(4), 839–850 (2009)

Robocup and Soccer Robots

Planning-Centered Architecture
for RoboCup SSPL @Home

Francisco Martín-Rico[1]([✉]), Jonathan Ginés[1], David Vargas[1],
Francisco J. Rodríguez-Lera[2], and Vicente Matellán-Olivera[2]

[1] Rey Juan Carlos University, 28943 Fuenlabrada, Spain
francisco.rico@urjc.es, {jgines,dvargas}@gsyc.urjc.es
[2] University of León, 24071 León, Spain
{fjrodl,vicente.matellan}@unileon.es

Abstract. Software architectures in robotics organize perception and
action to solve a problem. In this paper we present a description of the
software architecture applied by our team in the Social Standard Platform League (SSPL) of the @home category of RoboCup. This league
simulates a domestic scenario where the robot must interact with the
dependent people who live in it to help them in their daylife. This architecture is designed to solve the tests of this league, addressing the integration of navigation, interaction, generation of behaviors and perception.
Our architecture follows a three-layer organization where the core is a
classic planner that uses Planing Domain Definition Language (PPDL).
This architecture has been validated in the humanoid robot Pepper during our participation in the RoboCup 2018 in Montreal, Canada.

Keywords: RoboCup · Behavior architectures · Humanoid robot

1 Introduction

The RoboCup [1] is the most important robot competition in the world. Its goal
is that in the year 2050 the soccer human world champion will be defeated by
a team of robots. Although this competition, whose first edition was in 1997
in Nagoya [2], takes football as a reference, it has several leagues that present
different scenarios: rescue league, league @home and industrial league.

Presenting a specific problem focuses the efforts of the scientific community
in one direction. Artificial Intelligence had for decades an objective that made
advance its discipline: defeat the chess world champion. He achieved it in 1997
[3], an achievement that at the beginning of the challenge seemed impossible.
Like RoboCup, there are several competitions (RoCKiN [4], European Robotics
league [5], among others) that have made advance certain areas of research.
Currently, we are experiencing the emergence of commercial vehicles that drive
autonomously. The Grand Challenge [6] and the Urban Challenge [7] were competitions that a few years ago propitiated the advances to make this possible.

© Springer Nature Switzerland AG 2019
R. Fuentetaja Pizán et al. (Eds.): WAF 2018, AISC 855, pp. 287–302, 2019.
https://doi.org/10.1007/978-3-319-99885-5_20

The competitions have an added difficulty that has a beneficial effect on the development of the algorithms and software that run on the robots: the robot must work correctly at a certain time, without the option to repeat or synthetic tests. This brings a maturity and robustness to the systems that participate.

Our competition has as a fundamental characteristic that the used robot is not built by us, but rather it is a robot commercialized by Softbank. For this reason, the emphasis is not the making robots, but the programming of their system and their capabilities. This robot, in addition, is very limited in several issues. While it is a very attractive platform, with an ideal size and shape to interact with humans in domestic environments, its sensors and actuators (except the base) have very small capacities for a modern mobile robot. It has an 1.91 GHz Quad core processor with 4 GB DDR3 memory. It is equipped with two RGB cameras and one RGBD camera (with excessive distortion due to design issues) in the head. It has two sonar sensors and the laser sensors on its base. These laser sensors are actually three infrared cameras that generates information, very noisy and of short range, of distances, simulating a laser.

The software of a robot is a crucial component in a robot. As indicated in [8], it is important that robotics projects are not black holes in which the developments are born and die. It is important the developments survive the robots for which they were designed, or the projects that financed them. For make it happen, It is important to organize the software so that the modules that make up a robotic application are general and reusable between robots and projects.

This problem has been addressed in recent years by several robotic middleware, among which YARP and ROS [9] stand out. Both middlewares establish how to organize software into modules, and how communication occurs between them. Both have come to the conclusion that generalizing standard communication formats for each type of robotic data (images, laser, point cloud, etc.) allows the applications be independent of the drivers. Both middlewares are distributed, but the big difference is the amount of developers that form the community around each middleware. Although YARP offers a more effective communication system than ROS, the huge difference between software (drivers, libraries, tools, ...) makes ROS the standard in software development for robots.

The Pepper robot has its own middleware, called NaoQi [10]. From a software organization point of view in modules, it is similar to the middlewares described above. The disadvantage is that it is a real black hole, since it is only available for the development of the manufacturer's robots. For this reason we discard it beyond its use as a driver to access the sensors and actuators of the robot.

Middleware solves library availability, development and debugging tools, and software organization. This is not enough to address complex problems such as those proposed in the RoboCup. A Software Architecture is needed, which organizes perception and action to generate the behaviors that govern the operation of the robot.

Reactive approaches are not applicable to our complex problem. Deliberative approaches do not allow solving problems that depend on a continuous perception, because they calculate the complete sequence of actions, assuming an static world only updated by robot actions. In this paradigm, classical planning in Artificial Intelligence is widely used. The most standardized approach is to use Planning Definition Domain Language (PDDL) [11]. It is based on the codification of the problem that a robot or agent has to solve in two parts: the domain and the problem. The domain is the set of symbolic predicates, types it can handle and actions that can change the state. The actions have some prerequisites under which it can be applied, and some effects. A problem is a situation at a certain moment, made up of instances, concrete predicates, and a goal that must be obtained. The planners calculate the sequence of actions from the initial state (the problem to the goal). Classical planning, on its own, is not appropriate to solve complex tasks which needs continuous perception. Typically, it is integrated in more complex architectures, as we have done in this work.

The hybrid approaches are the most used for this type of problem. Rodney Brooks proposed the architecture of Subsuption [12], where the most complex behaviors were broken down into simpler behaviors. The simultaneous execution of these basic behaviors gave rise to an apparent intelligent behavior. This idea guided the development of BICA [13], which was used in robotic soccer. In BICA, the high-level components explicitly executed lower-level components, until they reached the components that obtained information from the sensors, or sent commands to the actuators.

For more complex tasks, architectures of more levels are usually used. In [14] a three-level architecture guided by motivational principles generated by the robot is proposed. The architecture is general in the medium and low levels, although the motivational system is not generalizable for general problems. In [15] also addresses a fault-tolerant architecture for robots to carry out long-term tasks, combining planning with Petri Nets [16]. Compared to our approach, it seems more complicated to adapt it for generalist tasks.

Our approach is built on top of ROS. It is a three-level architecture where the upper level reasons symbolically. It uses finite-state machines where states set goals and transitions evaluate predicates. The intermediate level calculates plans to obtain goals, and the lower level implements the actions, already in a concrete way analyzing the perceptions of the robot and sending commands to its actuators.

This paper is structured as follows: Sect. 2 describes our architecture of three levels centered on a planner, which is our main approach. Section 3 describes the general functionalities implemented in the architecture, capable of addressing general problems. Section 4 presents the competition tests, and how our architecture has been applied in its solution. Finally, Sect. 5 provides the final conclusions.

2   Software Architecture

Our software is based on ROS. We execute it on board the robot, in user space, since you can not modify the software already installed in the robot. ROS brings a very convenient middleware to develop applications for robots, as well as a huge amount of libraries, tools and software at our disposal. Still, ROS as middleware, does not provide a software architecture that organizes perception and action in a convenient way, so this must be implemented by us.

A good design of a software architecture for robots presents several questions to consider:

- There must be a good compromise between a development aimed at solving a specific problem and its ability to be reused for other applications.
- The architecture should not waste computing time, since the resources of a robot are limited.
- It must be easy to use. Not only when executing the software, but for the programmers of the different modules. Not all programmers should know in detail the operation of the core of the system to develop modules.
- It must be scalable. The growth in functionality or should compromise the operation or organization of the system.
- It must be robust and able to adapt to problems that may arise.
- It must make the behavior of the robot correct and vivacious.

For this reason, we have designed a three-layer architecture whose upper layer breaks down the problem into phases, allowing long-term behavior without overloading the others. The middle layer is implemented with a classic planning system that generates sequences of actions to obtain a goal. This layer provides the ability to recover from anomalous situations effectively. Finally, the lower layer implements the actions and basic capabilities of the robot.

The capabilities of the robot, perceptive or acting, must be available for use, but only at certain times. The rest of the time should not consume computing time. For this reason, we have applied the activation/deactivation mechanism of BICA, which specifies a convenient technique to save component computing power when execution is not needed.

2.1   Three-Layers Architecture

Initially, the architecture was divided into two levels instead of three. It was intended that the planning system solve the complete execution of long tasks that were composed of several phases, whose execution, in addition, could be bifurcated according to events during the execution. This had the disadvantage of getting the correct sequence of actions until achieving the final goal. In addition, a planning calculated the plan from the initial state to achieve the final goal. it was necessary to include facts that were not real, but allowed to reach the end, and trust that in the moments in which the real facts were established, force re-planning. It was complex to deal with the robot's tasks in this way. For this reason it was decided to include a top layer that divided the operation of the robot into phases, greatly simplifying the domain coding in PDDL.

Software architecture is shown in left part of Fig. 1. The top layer is a state machine, which can be hierarchical in turn, although we have not yet exploited

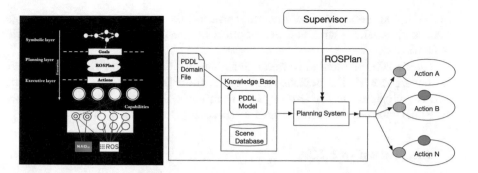

Fig. 1. Diagram of the three-layers Software Architecture (left) Planning System component (right).

this functionality. This layer only reasons at the symbolic level using predicates contained in the ROSPlan knowledge base, and establishing goals in each state, subsequently calling the planner.

The intermediate layer is the ROSPlan planning system, which generates, based on the goals established in the upper layer, a set of actions to satisfy it. The lower layer is formed by the components that carry out the actions that one by one the intermediate layer generates. These actions are the true implementers of the robot's functionality. Usually, these actions require the participation of perceptual abilities (detectors of people or objects, speech recognizer) or performance (voice generators, navigation module). These capabilities are in a pool of components that are activated when they are required using the mechanism described in the next section.

2.2 Planning System

Our software has as a core ROSPlan, which is a ROS module that allows calculating plans (i.e. a sequence of actions to satisfy an objective) from a general domain and a particular problem. ROSPlan is composed of two well differentiated parts: the knowledge base and the planning system (right part of Fig. 1):

- The knowledge base consists of two components: the pddl model, and a database that links symbolic instances with real data. For example, the pddl model could contain a location instance "kitchen", and the database could contain the coordinate where the kitchen is.
  The knowledge base can be updated and queried using ROS services. Once initialized from a file with the domain pddl, it is added, through these ROS services, the instances of the problem, the initial predicates and the goals.
- The planning system is activated by a service that triggers its execution. At that moment, he reads the information from the knowledge base, creates a plan, and dispatches the actions to carry it out. ROSPlan provides an interface to implement the actions that have been defined in the PDDL model. The planning system delivers the action, and waits for it to be completed. If the

action is completed successfully, add the effects of the action to the knowledge base and continue with the next action in the plan. If the action fails, its execution ends.

The actions implemented to manage the output of the Planning System must be coherent with the actions declared in the PDDL Domain File. Different project which use ROSPLan requires their corresponding actions. Any change in PDDL domain implies changes in these implementation of actions

2.3   Integration of BICA

BICA [13] (Behavior-based Iterative Component Architecture) was a software architecture used in the competition of robotic soccer. This architecture defined the behavior of the robot as the iterative execution of components that formed an activation tree. The components of the root of the tree defined at a high level the behavior of the robot, many implemented as finite state machines, and the leaves of the tree were perceptive or acting components. BICA had a very attractive property: the components activated other components. Even each state of a finite state machine implemented within a component activated their required components. BICA had a sophisticated system by which the components declared their dependencies (the components that it activated), and kept track of the components that activated it. If one of the components had at least one active component that activates it, it was executed at a fixed frequency, defined locally by the component.

BICA evolved from its original implementation towards ROS. Now the BICA components are ROS nodes (BICAROS node), preserving their characteristics. Multiple BICAROS nodes can be launched. A node is inactive until at least one component that has declared it as dependency becomes active.

In our architecture, all capabilities are BICAROS nodes. In such a way that every one is inactive until any BICAROS node activates it. The nodes that implement the actions defined in the PDDL domain are BICAROS nodes that are activated when they are required by the planner, activating in cascade the capacities that it needs to carry out its work. This mechanism greatly simplifies the development of our architecture software, increasing the scalability of the system.

3   Robot Functionality

Once the software architecture is described, in this section we will describe how we have used it to implement the robot's functionalities.

To carry out the tests of the RoboCup, which functionalities are really what a home assistant robot should have, we have implemented certain basic functionalities. In this section we will describe the most important: the robot's ability to move around a domestic environment, the perception of objects and people, and multimodal interaction with people.

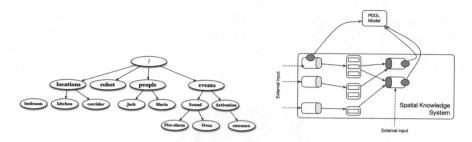

Fig. 2. Example of knowledge taxonomy (left). Spatial knowledge system (right).

3.1 Perception

Normally, each source of information is processed separately, being complicated to merge it to obtain added knowledge. We use an octomaps-based common representation for all informations with spatial meaning, organized in taxonomies. This representation allows to represent probabilistically in the space any information that the robot may need, as well as to generate new information by applying simple operators between these octomaps. Our approach is integrated into a planning system based on PDDL where the predicates are generated and dynamically updated to be aware of a changing context.

We propose grouping these knowledge into taxonomies with a hierarchical structure, as shown in left part of Fig. 2. The knowledge is represented as a tree, where the leaves are an instance that is represented by a different octomap: The existence of `/people/Jack` already indicates that there is a person called Jack, and his associated octomap indicates the probability of where it is. Also, the last time this octomap was updated could represent the last time it was perceived. Similarly, `/locations/kitchen` indicates where the kitchen is and `/robot` where the robot is.

Data from different taxonomies can be combined to produce more elaborate information. Combining the data of people and places could produce the knowledge of where each person is. This would be done with simple intersection options between octomaps. Combining the data of objects could produce information about the relative position of objects, with simple operations of distances and positions between octomaps ("close to", "above", ...).

The spatial knowledge system generates or eliminates predicates for the knowledge base, based on the spatial relationships of the different instances perceived by the robot or any knowledge it has. This allows the knowledge base to be coherent with the reality perceived in real time. In the right part of Fig. 2 is shown the diagram of this system. The system is composed of two types of components: *producers* and *consumers*. The former communicate with the latter in an N:N decoupled communication scheme that fits with the publication/subscription paradigm.

Producers produce two types of information: (i) Instances for the knowledge base, and (ii) octomaps, one for each instance, representing its position/probability in the space. The number of instances can be inited from files, o it can be perceived by sensors.

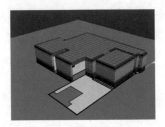

Fig. 3. Octomaps corresponding to locations in a house. Each room is represented with an octomap. Colors only represent Z-axis and bedroom is deactivated in this figure to show that there are many individual octomaps, one for each room.

Figure 3 shows an example of the output of one of these spatial information producers. This producer reads a set of location (kitchen, corridor, bedroom, ...) in a house from a file containing their dimensions and position. For each location, this component:

- Adds an location instance to the knowledge base.
- Adds the representing coordinates to the database, connecting it to the symbolic instance of the location (for example, *kitchen*).
- Publishes an octomap to represent the space it occupies.

Consumers are responsible for extracting relationships between octomaps to update predicates in the knowledge base. A consumer can subscribe to several types of spatial information.

Let's see some examples:

- A consumer could subscribe to the information of people to generate predicates such as (person_near Jack Nick).
- A consumer could subscribe to the information of the people and to the locations to generate predicates such as (person_at Jack Kitchen). To do so, I would only have to check the spatial intersection of each octomap of people with each octomap of locations and when they overlap, generate the knowledge.
- A consumer could subscribe to the information of the sound events and to the locations to generate predicates such as (sound_at Doorbell corridor).
- A consumer could subscribe to the information of the locations and the information, external to the architecture, of the position of the robot generated by its self-localization module to generate predicates in real time such as (robot_at kitchen).

These components are also responsible for removing from the knowledge base those predicates that detect false. If detected (robot_at kitchen), any other predicate (robot_at X) is removed, if X is not kitchen.

3.2   Navigation

Navigating in indoor environments also involve complex actions, such as opening doors, using elevators, and many others. We use a topological navigation system based on AI Planning. Starting from a PDDL representation of the environment as predicates, navigation tasks are divided into phases, in which different actions are required. This approach has proven to be very effective to plan the operations of a robot, including navigation, at indoor environments.

A topological map divides the environment into positions that have meaning from the point of view of the actions that can be performed on it. In the upper left corner of Fig. 4 a small coded environment is shown as an occupation map. In the upper part of the right we have identified the areas where the robot must initiate complex actions, such as of crossing a door, or open it or ask to open it if it is closed. In the middle part of this Figure we have been identified nodes that correspond to each position, taking into account that in each position the robot can be arranged to cross the door, or is the point where the action of crossing it ends. Finally, the bottom part of the figure represents the transitions between the different nodes. Each different color in the transitions indicate different actions that must be carried out to move from one node to another, as we will see below.

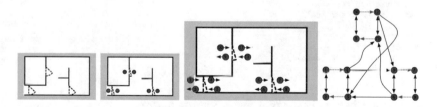

Fig. 4. (Up) Occupancy maps with doors position indicated. Red circles represent critical positions where actions must be taken. (Middle) decomposition in different nodes with orientation, and (Bottom) their representation in a topological map as a directed graph.

The PDDL domain contains the actions that the robot can perform to solve a plan. In our case, the actions are associated to the different types of transitions between the nodes of the topological map. Action Move performs navigation among waypoints in the same room, and action Cross performs navigation crossing the door. Both actions use the ROS navigation module, but in Cross the robot waits until a door is opened (asking for it if necessary) or even marks a door as closed.

The topological map is encoded in the form of PDDL predicates. The domain has the following predicates:

```
(:types
    location
    robot
)
(:predicates
    (at ?r - robot ?l - location)
    (connected ?from ?to - location)
    (door ?from ?to - location)
)
```

The PDDL domain contains the actions that the robot can perform to solve a plan. In our case, the actions are associated to the different types of transitions between the nodes of the topological map:

```
(:durative-action move
  :parameters (?r - robot ?from ?to - location)
  :duration ( = ?duration 10)
  :condition (and
    (at start(at ?r ?from))
    (over all(connected ?from ?to))
  )
  :effect (and
    (at end(at ?r ?to))
    (at start(not (at ?r ?from)))
  )
)

(:durative-action cross
  :parameters (?r - robot ?from ?to - location)
  :duration ( = ?duration 20)
  :condition (and
    (at start(at ?r ?from))
    (over all(door ?from ?to))
  )
  :effect (and
    (at end(at ?r ?to))
    (at start(not (at ?r ?from)))
  )
)
```

Each action has its corresponding *action executor*.

When the topological map manager receives a request to navigate to a point on the map, the navigation process begins. First, updating the topological map as described in the previous section. Secondly, setting the node R to the position of the robot with the information of self-location of the robot. Finally, setting the goal, indicating that node G is the position to which the robot must navigate:

```
(robot leIA)
(:goal (at leIA wp_goal))
```

3.3   Human-Robot Interaction

The interaction with human beings in the main characteristic of a social robot. For this reason, the interaction must be done naturally through the voice, mainly. In addition, this interaction should not produce frustration, and this can happen if the robot does not understand people correctly, or if the robot is not able to speak in a tone acceptable to anyone.

Interaction is key during RoboCup tests. If there is a lot of noise, or there are network problems that prevent us from using web services for this, very common, the interaction must occur by any alternative means: *show must go on.*

For these reasons we have implemented different redundant interaction mechanisms that are used in the absence or incapacity of others, as shown in Fig. 5.

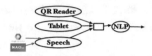

Fig. 5. HRI system

4   Discussion

Performing a complete and exhaustive experimentation of a software architecture is a complex task and, in many cases, not approachable. In fact, robotic competitions provide a means of evaluating complete robotic systems through the tests it poses:

- The variety and difficulty of the tests demonstrates the versatility of the proposed solutions. The system is sufficiently broad and scalable to deal with different problems.
- The capacities that are required in each test indicate the effectiveness in integrating the perception and action required in each one.
- Robustness is intrinsic to competition. The robot does not have all the opportunities it could have in a laboratory test. At the start of the test, the robot must be on the starting line, regardless of the conditions of the network, the scenario, or any other; the robot must operate without delays or retries.

The Standard Service Platform League has as its main characteristic that all the participants use the same robot, focusing their efforts towards software development. A robot cannot be modified in any way: a robot cannot be equipped with more processing units, or better sensors. External computing is allowed. The competition offers wireless connection with Internet access, but does not guarantee that it works, nor the quality of the connection. For this reason, it is necessary to implement redundant and alternative methods in case of not having external computing, in case of needing it, and to comply with one of the main rules of the competition: Show must go on. For example, we use Google Speech to understand the voice commands, but we have implemented alternative means of man-robot interface, as described in Sect. 3.3.

For these reasons we will describe the tests in which the architecture was applied. Our architecture successfully faced these tests, showing its main capabilities. Each test has a set of points that are awarded if it correctly fulfills the tasks it must perform, partially scoring if it is not completely satisfied.

4.1   Help Me Carry

In this test the robot starts at a known place in the arena (the scenario that simulates a house). The referee starts in front of the robot. The robot instructs the referee to make it easier for the robot to learn the referee's appearance. Once this phase is finished, the robot must follow the referee out of the house, to a place not known a priori. When the referee decides, he tells the robot to go to a specific room in the house to look for help. The robot must return to the house and, in the room indicated by the referee, ask a person to follow him to where the referee was.

To overcome this test, we have developed three main actions:

- **Learn to the referee**. This action is based on the ability to extract characteristics of people (man/woman, color of clothes, hair color, ...) perceived by his camera. This action lasts until all these characteristics have been perceived.
- **Follow a person**. This action is based on two capacities: The capacity described in the previous point, and the ability to trace a route to the person who meets the characteristics of the referee. For this reactive navigation capability, a local navigation algorithm based on Gradient Path Planning has been developed, using the laser to detect obstacles and thus trace the route, as shown in right part of Fig. 6.
- **Guide a person**. This action uses the back sonar to navigate to the referee's position while checking that it is behind. At any time, the guided person can stop following the robot, and the robot must realize it.

All the test must use the interaction of the robot, both to understand the commands and to give instructions to humans. In addition, the navigation system must have SLAM capabilities, to be able to navigate to unknown areas, and be able to return. For this we have used the navigation system described in Sect. 3.2.

4.2   Speech and Person Recognition

In this test, the robot starts outside the house. When the door is opened, the robot goes to the living room, with its back to a sofa (left part Fig. 6). When the referee indicates, the robot turns around and has to say all the available information about a group of people on the sofa: number, how many men and how many women, their clothes, ... Once the result is communicated of this recognition, several people are placed around the robot and begin a game of questions and answers, which includes 3 questions among the topics: general (from a bank of questions known a priori), questions about elements of the arena and questions about their team. At the end, there are 3 more questions, in which the robot must detect the source of the sound and turn to it before answering.

For the question and answer game we have used a pre-processed question-answer bank with the output of the NLP module. When a question is asked, it is processed by the NLP module and the distance to the existing questions in the database is calculated. This distance punctuates the similarities (taking into account synonyms) depending on whether they are names or verbs (these are more important for calculating distance), or other elements such as prepositions or determinants. If you exceed a threshold, the stored response is offered. It may seem a trivial method, but it is effective.

Fig. 6. Speech and Person Recognition test during the trial of our team (left). Gradient Path Planning implementation for the Follow Person action (right).

4.3   Cocktail Party

In this test, the robot starts outside the house. When the referee opens the door, the robot navigates to a point in the dining room. There are several groups of people there. Some of them call the robot to come and take note of what they want to take. The robot must collect the name and characteristics of each client. Once all the requests are collected, the robot goes to where the bartender is (a priori known position) and tells him what each person wants, giving a precise description of each one. The bartender can indicate that a drink is not available, and the alternatives. The robot must return to where the client is (which may have changed position) and offers alternatives. After knowing the client's option, he returns to the bartender and communicates it to him.

In this test, the main difficulty is to identify when a person calls the robot, as well as to perceive its characteristics. Interaction is a challenge, in which the NLP system is key. In the dialog, we must take into account the set of valid requests given the context of the conversation. Knowing the set of beverages available, the task is facilitated. In addition, the confirmation system and an alternative interface on the tablet (where you can select beverages) allow you to save requests that are not understood in a first attempt.

4.4   General Purpose Service Robot

This is a complex test, since the set of tasks that the robot can face is very large. The robot starts from a pre-established position inside the arena. The referee asks you to perform three missions. Examples of each of the missions:

```
"Go to the bedroom, find Jack, and tell him what is the name of your team"
"Go to the living room and tell Lorena how many apples are on the table"
```

The set of orders is generated by an application available on the competition website. It contains many orders arranged in three categories of difficulty. Our solution is similar to that of the test described in the Sect. 4.2. We have a database that associates each order with one or more goals in PDDL. From an entry order, we obtain the closest order using a distance based on weights and the classification of the words, and return one or more corresponding goals.

When we receive a mission, we divide it into orders using the NLP module. We obtain the goals that correspond to each order in the database. We divide the sentences of the missions into specific orders. We process them in order and we queue them together. The robot performs them in order and returns when he has finished the next one, or leaves the arena if he has already completed all three missions.

5   Conclusions

In this paper we have presented a software architecture for mobile robots, capable of carrying out complex and diverse tasks. The main focus of its design has been the facility to incorporate new functionalities, and the ability to adapt to environmental situations.

This architecture is divided into 3 levels. The upper level handles the symbolic domain, being implemented as state machines. The main objective of this layer is to establish the goals of which each mission is composed by the robot. The intermediate layer calculates the sequence of actions to carry out a goal, being able to adapt to the conditions of the environment. The lower layer implements the actions that the robot can carry out, as well as the capabilities it may require.

Within this architecture we have implemented a topological navigation system, capable of dividing robot navigation into actions such as moving around a room or crossing doors, and can be extended to actions such as crossing security lathes or taking elevators. We have also implemented a system of spatial perception that generates symbolic knowledge, coding the probability of occupation of a certain object or person in a hierarchy of octomaps. Finally, we have implemented a man-machine communication system with fault-tolerant redundant input.

This architecture has allowed us to address all the tests during our participation in the RoboCup 2018 in Montreal, in the category of the Social Standard Platform League. In this league a set of tests are presented where the capabilities that a social robot must have in a domestic environment must be applied. We have presented the tests to show the complexity of tasks that our architecture can deal with.

Acknowledgments. The research leading to these results has received funding from the RoboCity2030-III-CM project (Robótica aplicada a la mejora de la calidad de vida de los ciudadanos. fase III; S2013/MIT-2748), funded by Programas de Actividades I+D en la Comunidad de Madrid and cofunded by Structural Funds of the EU. It has also received funding from the RETOGAR project (TIN2016-76515-R) from the

Spanish Ministerio de Economía y Competitividad. This research has received material from NVidia Grants Program.

References

1. Burkhard, H.-D., Asada, M., Bonarini, A., Jacoff, A., Nardi, D., Riedmiller, M.A., Sammut, C., Sklar, E., Veloso, M.M.: RoboCup - yesterday, today, and tomorrow. workshop of the executive committee in Blaubeuren. In: Robot Soccer World Cup (RoboCup), October 2003, pp. 15–34 (2003). https://doi.org/10.1007/978-3-540-25940-4_2
2. Kitano, H., Asada, M., Kuniyoshi, Y., Noda, I., Osawa, E.: RoboCup: the robot world cup initiative. In: Proceedings of the First International Conference on Autonomous Agents (AGENTS 2097), pp. 340–347. ACM, New York (1997). https://doi.org/10.1145/267658.267738
3. Campbell, M., Hoane, A.J., Hsu, F.: Deep blue. Artif. Intell. **134**(12), 57–83 (2002). https://doi.org/10.1016/S0004-3702(01)00129-1
4. Amigoni, F., Bonarini, A., Fontana, G., Matteucci, M., Schiaffonati, V.: 2nd Workshop on Robot Competitions: Benchmarking, Technology Transfer, and Education, European Robotics Forum 2013, Lyon (2013). https://doi.org/10.5772/intechopen.69115
5. Lima, P.U., Nardi, D., Kraetzschmar, G.K., Bischoff, R., Matteucci, M.: RoCKIn and the European robotics league: building on RoboCup best practices to promote robot competitions in Europe. In: Behnke S., Sheh R., Sarel S., Lee D. (eds.) RoboCup 2016: Robot World Cup XX. RoboCup 2016. LNCS, vol 9776. Springer, Cham (2016). https://doi.org/10.1007/978-3-319-68792-6_15
6. Buehler, M., Iagnemma, K., Singh, S.: The 2005 DARPA Grand Challenge: The Great Robot Race, 1st edn. Springer (2007, incorporated)
7. Buehler, M., Iagnemma, K., Singh, S.: The DARPA Urban Challenge: Autonomous Vehicles in City Traffic, 1st edn. Springer (2009, incorporated)
8. Fitzpatrick, P., Metta, G., Natale, L.: Towards long-lived robot genes. Robot. Auton. Syst. **56**(1), 29–45 (2008). https://doi.org/10.1016/j.robot.2007.09.014
9. Quigley, M., Conley, K., Gerkey, B.P., Faust, J., Foote, T., Leibs, J., Wheeler, R., Ng, A.Y.: ROS: an open-source robot operating system. In: ICRA Workshop on Open Source Software (2009)
10. Naoqi Development Guide. http://doc.aldebaran.com/2-5/index_dev_guide.html
11. Mcdermott, D., Ghallab, M., Howe, A., Knoblock, C., Ram, A., Veloso, M., Weld, D., Wilkins, D.: PDDL - the planning domain definition language. Technical report, CVC TR-98-003/DCS TR-1165, Yale Center for Computational Vision and Control (1998)
12. Brooks, R.A.: Elephants don't play chess. Robot. Auton. Syst. **6**(12), 3–15 (1990). https://doi.org/10.1016/S0921-8890(05)80025-9
13. Aguero, C.E., Canas, J.M., Martin, F., Perdices, E.: Behavior-based iterative component architecture for soccer applications with the NAO humanoid. In: Proceedings of the 5th Workshop on Humanoid SoccerRobots @ Humanoids 2010, Nashville, pp. 29–34 (2010)
14. Rodríguez-Lera, F.J., Matellán-Olivera, V., Conde-González, M., Martín-Rico, F.: HiMoP: a three-component architecture to create more human-acceptable social-assistive robots: motivational architecture for assistive robots. J. Cogn. Process. **19**(2), 233–244 (2018). https://doi.org/10.1007/s10339-017-0850-5

15. Iocchi, L., Jeanpierre, L., Lázaro, M.T., Mouaddib, A.-I.: A practical framework for robust decision-theoretic planning and execution for service robots. In: Proceedings of the Twenty-Sixth International Conference on Automated Planning and Scheduling (ICAPS 2016), pp. 486–494 (2016)
16. Murata, T.: Petri nets: properties, analysis and applications. Proc. IEEE **77**(4), 541–580 (1989). https://doi.org/10.1109/5.24143

Opponent Modeling in RoboCup Soccer Simulation

José Antonio Iglesias$^{(\boxtimes)}$, Agapito Ledezma, and Araceli Sanchis

Universidad Carlos III de Madrid, Avda. de la Universidad,
30, 28911 Leganés, Madrid, Spain
{jiglesia, ledezma, masm}@inf.uc3m.es

Abstract. RoboCup is an international scientific initiative with the goal to advance the state of the art of intelligent robots. RoboCup offers an integrated research task covering broad areas of Artificial Intelligence and robotics. Within this competition and without the necessity to maintain any robot hardware, the RoboCup Simulation League is focused on artificial intelligence and team strategy. This league can be considered as a multi-agent domain with adversarial and cooperative agents where the team agents should be adaptive to the current environment and opponent. In this paper, we present an approach for creating and recognizing automatically the behavior of a simulated soccer team.

Keywords: RoboSoccer · Simulated soccer team · Opponent modelling

1 Introduction

Over the last two decades, Artificial Intelligence (AI) techniques have been used successfully in many areas such as science, engineering, business, medicine or weather forecasting. However, to encourage progress in AI research, it is useful to solve different and difficult benchmark problems. In this sense, RoboCup offers, from 1997, an important set of these benchmark problems in the framework of an international competition [1]. RoboCup is an international scientific initiative with the goal to advance the state of the art of intelligent robots. RoboCup offers an integrated research task covering broad areas of Artificial Intelligence and robotics. Such areas include: real-time sensor fusion, reactive behavior, strategy acquisition, learning, real-time planning, multiagent systems, context recognition, vision, strategic decision-making, motor control, intelligent robot control, and many more [2].

Within this competition and without the necessity to maintain any robot hardware, the RoboCup Simulation League is focused on artificial intelligence and team strategy [3]. This league can be considered as a multi-agent domain with adversarial and cooperative agents where the team agents should be adaptive to the current environment and opponent [4]. This league introduces an online method to provide the agents with team plans that "coach" agent generates in response to the specific opponents. The coach agent is equipped with several predefined opponent models.

Considering the dynamical and adversarial nature of a soccer play, opponent modeling is very relevant in the RoboCup environment, especially in the simulation league. For this reason, a new competition was created in 2001: *RoboCup Coach*

© Springer Nature Switzerland AG 2019
R. Fuentetaja Pizán et al. (Eds.): WAF 2018, AISC 855, pp. 303–316, 2019.
https://doi.org/10.1007/978-3-319-99885-5_21

Competition, in which an on-line coach was able to act as an advice-giving agent [5]. The main goal of this competition was to emphasize opponent-modeling approaches, and it was held every year (with some changes) from 2001 to 2006. However, several considerations were not well defined, and the competition was suspended after RoboCup Coach Competition 2006. In [6] a new approach for that competition was proposed to face the opponent modeling challenge.

However, although the competition was suspended, the platform and the idea behind that competition have continued to be used and it sparked a growing interest in the area of agent/team modeling [7–10].

In this paper and considering the idea behind the *RoboCup Coach Competition 2006*, we present an approach for creating and recognizing automatically the behavior of a simulated soccer team. The paper is organized as follows: First, the related work is presented, specially the previous research done by the authors of this papers. Then, the structure of the *RoboCup Coach Competition 2006* is detailed. After this, our approach is presented and evaluated. Finally, some concluding remarks are presented.

2   Related Work

An opponent model can be defined in general as an abstracted description of the behavior of one or several players in a game. The beginning of opponent modeling is a work done in 1996 by Carmel and Markovitch [11], in which it is introduced the M* algorithm, a generalization of the minimax algorithm that uses an arbitrary opponent model to benefit from its flaws. In [12] it is assumed that the strategies of the agents can be modeled as finite automate and a model-based approach is presented as a method for learning effective interactive strategies.

There are many dynamic multiagent domains with adversary agents in which opponent can be implemented efficiently. In this paper, we focus the opponent modeling task in (simulated) soccer. Soccer a competitive and collective sport in which teammates execute basic actions to lead their team to more advantageous situations. In this sport, if a team can recognize and extract the opponent modeling, it can be very useful to achieve the final goal.

In [13] an approach to perform post-hoc analysis of RoboCup Soccer Simulation 3D teams via log files of their matches is presented. In addition, the approach can learn a model to classify the teams not only as being strong, medium or weak but also through their game playing styles such as frequent kickers, frequent dribblers, heavy attackers, etc. Using the learned model, the teams can be clustered to predict game style of similar opponents.

In [8] a methodology based on the use of a plan definition language to abstract the representation of relevant behaviors is proposed. Experiments were conducted based on a set of game log files generated by the Soccer Server simulator which supports the RoboCup 2D simulated robotic soccer league. The effectiveness of the proposed approach was verified by focusing primarily on the analysis of behaviors which started from set-pieces and led to the scoring of goals while the ball possession was kept.

In [14], the authors present neural-based models that jointly learn a policy and the behavior of opponents. These models are inspired by the recent success of deep reinforcement learning. In this work, the authors encode observation of the opponents into a deep Q-Network (DQN) and retain explicit modeling using multitasking. By using a Mixture-of-Experts architecture, their model automatically discovers different strategy patterns of opponents without extra supervision.

In relation to the topic of this research, the authors of this paper have already developed several researches works:

In [15] we described the main features of the team *Caos Coach 2006 Simulation Team* which participates in that competition. That description team paper focuses on the challenge of the opponent modelling. This paper is based on that description but in this case, the detection of patterns is different, and the method is widely evaluated.

In addition, in [16] the authors presented a tool, called *Viena,* which can visualize a *RoboCup Soccer Game* using a dynamic data representation, detect soccer actions and analyze a soccer team behavior by using different algorithms creating a soccer team behavior model.

Finally, in a previous work [17], we proposed a novel approach to use efficiently the opponent models detected to improve the behavior of a team. For using these models, it was necessary a special agent (*coach*) which can model the observed opponent team and communicate a counter-strategy to the coached players. However, in the present research, we are modelling the opponent and detecting their play patterns, but not *using* the detected models.

3   RoboCup Coach Competition

The RoboCup Coach Competition structure has been changing gradually from 2001 to 2006. In this section, we describe in detail the structure of the competition that we consider in this paper in the opponent modeling task: *RoboCup Coach Competition 2006.* This competition is also explained in [6].

Before describing the structure of this competition, it is necessary to define the following two concepts:

- *Play Pattern*: Simple behavior that a team performs which can be predictable and exploitable. In this paper we use the term pattern as a contraction of play pattern.
- *Base Strategy*: The general strategy of a team regardless of the pattern in it.

According to the RoboCup 2006 Coach Competition official rules, previously to the competition, a set of strategies are used as the base strategies of the patterns which are created by the organizing committee and some games are played (*no-pattern log files*). Then, the patterns are added to these base strategies, and some sample games are played again (*pattern log files*). Many pairs of log-files (*pattern log file* and its corresponding *no-pattern log file*) are created.

At the beginning of the competition, each coach team participant is provided with some pattern game log-files (only one pattern is activated in a log-file) and its corresponding *no pattern game log-files* (games with the same base strategy but the pattern not activated). The main goal in this competition is to look for the qualitative differences among the *pattern log file* and its corresponding *no pattern log file*. The coach should detect the patterns followed by the test team in the *pattern log files* and report

them. Also, once every pattern has been detected and stored, the coach should recognize them by observing a live game.

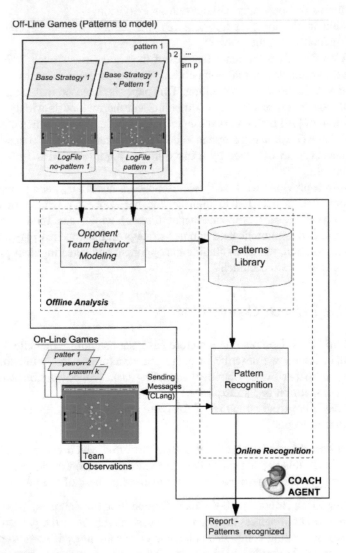

Fig. 1. Overview structure of the *RoboCup Coach Competition*.

Thus, this competition consists of two phases:

- *Offline Analysis:* The inputs of this phase are several pattern log fules and its
  corresponding *no-pattern log fules*. The coach must look for the qualitative differences between the two log files to detect the pattern. The output of this analysis is
  a set of files (*Pattern Library*) where the specifications of all the patterns are
  recorded.

- *Online Recognition:* The coach observes a live game where some patterns have been activated in the test team. The coach should recognize on-line the patterns activated in the test team in a 6000 cycles game and report them. The sooner the coach sends the report, the more score it gets.

Figure 1 shows the overview structure of the RoboCup Coach Competition. We can observe that the research is focused on team/opponent modeling and on-line recognition since the performance of a given coach is based only on its ability to detect and report patterns.

4 Our Approach

The approach proposed in this paper is based on a novel method developed by the authors of this paper for creating and recognizing automatically the behavior profiles of users from the commands they type in a command-line interface [18]. In the RoboCup environment, a previous work [15] proposes a different approach (especially in the online analysis) for the goal proposed in this paper.

In this research, we consider that the pattern of a team can be represented by a sequence of a high-level soccer actions. Thus, the first step in our approach is the detection of high-level actions (Sect. 4.1). Then, the two phases of the competition can be tackled by the proposed method (Sects. 4.2 and 4.3).

4.1 Detection of Soccer Actions

This task has also been detailed in a research in which a tool, called *Viena*, was presented [16]. Previously, we also used this idea in [15].

To detect the high-level soccer actions during a soccer game, it is necessary to extract the important information from the *logfiles*. With this purpose, in [19] is described a procedure to identify high-level actions in a soccer game, where an action represents a recognized atomic behavior. Based on that work we created the following method to detect soccer actions from the *logfiles* of a soccer game.

At every cycle of the server, each agent updates its world model with the most relevant data of these *logfiles*. In every cycle the following data is extracted:

- *Cycle*: Number that enables arrange the actions.
- *Ball Position*: Represented by a point in the Cartesian coordinate system.
- *Teammate Positions* and *Opponent Position*: Each player position is represented by a point in the Cartesian coordinate system.
- *Ball Possessor*: Number that indicates who is the owner of the ball.

After extracting these data from the *logfiles*, the actions that have occurred must be inferred. We need to consider that some actions are very hardly to identify, even if it is done by a soccer expert. Considering the work mentioned above [19], we propose to identify the following nine different actions.

- PassXtoY: Player X of the team T kicks the ball and a teammate Y gains possession, then the ball owner made a pass. Perhaps the ball owner did not want to do this
  pass, but we cannot consider this assumption.
- InterceptedPassXtoY: Player X kicks the ball and the opponent Y gains possession
  within a reasonable distance of the ball owner, the action is assumed to be an
  intercepted pass.
- StealXfromY: Player X kicks the ball and an opponent Y gains possession, then the
  opponent stole the ball from the ball owner.
- FoulX: Player X commits a foul.
- ShootX: Player X kicks the ball and at the end of the interval, the ball is close to the
  goal.
- GoalX: Player X kicks the ball and at the end of the interval, the ball is in the goal,
  the action is classified as a goal by the player X.
- KickInX: Player X kicks the ball when the state of the game is "kickIn".
- KickOffX: Player X kicks the ball when the state of the game is "kickOff".
- CornerX: Player X kicks the ball after other player shoot the ball.
- Free Kick: The goalkeeper of the team T kicks the ball after holding it.

Once the soccer actions are obtained, they are ordered creating a sequence of events
that represent the behavior of a specific team during a game. To differentiate from the
two teams, each sequence of a team soccer actions is enlarged while that team owns the
ball. Thus, a sequence of soccer actions represents the behavior done by a soccer team
while that team owns the ball, so a game will be represented by many sequences of
actions.

Let us consider the following sequence of actions as example:

{Pass1to2 → Pass2to1 → Pass1to2 → Pass2to1 → Shoot1}

4.2   Offline Analysis

In this phase, the first step is to extract the significant pieces of the sequence of soccer
actions that can represent a play pattern. In this sense, as it was used in [20], to get the
most representative set of subsequences from the acquired sequence, the use of a *trie*
data structure [21] is proposed. This goal is done by segmenting the sequence of action,
storing the corresponding subsequences in a *trie*, and then, creating the behavior team
model. This process is explained in the following 3 subsections. To clarify this process,
it is used as example the sequence presented in the previous subsection.

Segmentation of the Sequence of Soccer Actions

Each sequence is segmented in subsequence of equal length from the first to the last
action. Thus, the sequence $A = A_1 A_2 \ldots A_n$ (where n is the number of actions of the
sequence)   will   be   segmented   in   the   subsequences   described   by
$A_i \ldots A_{i+length} \forall\, i, i = [1, n - length + 1]$, where length is the size of the subsequences
created and determines how many actions are considered as dependent. In the rest of
the paper, we will use the term subsequence length to denote the value of this length. In
the proposed sample sequence *{Pass1to2 → Pass2to1 → Pass1to2 → Pass2to1 →
Shoot1}*, let 3 be the subsequence length, then the following 3 subsequences are
obtained:

{Pass1to2 → Pass2to1 → Pass1to2}, {Pass2to1 → Pass1to2 → Pass2to1} and {Pass1to2 → Pass2to1 → Shoot1}.

Storage of the Subsequences in a Trie

The sequences of soccer actions are stored in a *trie* in a way that all possible sequences are accessible and explicitly represented. In the proposed *trie*, a node represents a soccer action, and its children represent the soccer actions executed by the team after that. Also, each node keeps track of the number of times an action has been inserted on to it. As the dependencies of the commands are relevant in the team behavior, the subsequence suffixes (subsequences that extend to the end of the given sequence) are also inserted.

Considering the previous example, the first subsequence *{Pass1to2 → Pass2-to1 → Pass1to2}* is added as the first branch of the empty trie (Fig. 2a). Each node is labeled with the number 1 (in square brackets) which indicates that the soccer action has been inserted in the node once. Then, the suffixes of the subsequence (*{Pass2-to1 → Pass1to2}* and *{Pass1to2}*) are also inserted (Fig. 2b). Finally, after inserting the 3 subsequences and its corresponding suffixes, the completed trie is obtained (Fig. 2c).

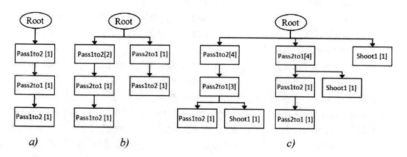

Fig. 2. Creation of the corresponding *trie* - example.

Creation of the Team Behavior Model

For this purpose, frequency-based methods are used. Specifically, to evaluate the relevance of a subsequence, its relative frequency or support [22] is calculated. In this case, the support of a subsequence is defined as the ratio of the number of times the subsequence has been inserted into the *trie* to the total number of subsequences of equal size inserted. Calculating this value, the *trie* is transformed into a set of subsequences labeled with its corresponding support value. This structure is represented as a distribution of relevant subsequences. Once a team behavior model has been created, it is stored in the *Patterns Library* with an identification name. In the previous example, the *trie* consists of 9 nodes; therefore, the profile consists of 9 different subsequences which are labeled with its support (Fig. 3).

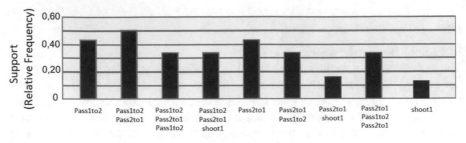

Subsequences of soccer actions

Fig. 3. Distribution of subsequences of soccer actions - Example.

4.3 Online Analysis

In this phase, a new game is observed, and the patterns followed by the opponent team need to be recognized. In this environment, the behavior team model is obtained and compared with the models stored in *Pattern Library*. It means that observing a team (T) and considering a set of team patterns $P = \{tp_1, tp_2, \ldots, tp_n\}$ stored in the *Pattern Library*, it is needed to determine which pattern $tp_i \in P$ are the observed team (T) following.

In a previous work [15] this phase was done by creating a new *trie* by observing the *new* team and compare it with all the previous tries already stored. In that case an algorithm to compare two tries is proposed.

However, in this research, we propose not to compare the tries, but the distributions. In this sense, the distribution of relevant actions of the observed team is created by applying the process explained in the previous section. Then, it is matched with all the patterns stored in the *Pattern Library*. As both team behaviors are represented by a distribution of values, a statistical test is applied for matching these distributions. A non-parametric test (or distribution-free) is used because this kind of tests does not assume a particular population distribution. In this case, it is used a modification of the Chi-Square Test for two samples. This method was also applied in [18] with the same goal but in a very different environment.

To apply the proposed test, the observed sequence is considered as an observed sample and the patterns stored in the *Pattern Library* are considered as the expected samples. Then, this test compares the observed distribution with all the expected distributions objectively and evaluates whether a deviation appears. The Chi-Square Test compares the two sets of support values in which *ChiSquare* is the sum of the terms: $\chi^2 = \frac{(Obs-Exp)^2}{Exp}$ calculated from the observed (*Obs*) and expected (*Exp*) distributions.

However, using this test, all the expected values are compared but if an observed value is not represented in the expected distribution, it is not considered. For this reason, the way to compare the two distributions is modified to the sum of the terms (this value is named as χ^2_Obs) as follows: $\chi^2\_Obs = \frac{(Exp-Obs)^2}{Obs}$. Using this test, a value that indicates the deviation between the observed and the stored profile is obtained. This deviation needs to be calculated with all the patterns stored and the observed team behavior that obtains the lowest deviation value indicates the closer similarity. In addition, the degrees of freedom (*dof*) are the same in all the comparisons with the expected patterns.

As example, let us consider that the previous example sequence represents the observed team behavior which follows several patterns: {*Pass1to2* → *Pass2-to1* → *Pass1to2*}. In addition, the pattern library contains already two different patterns ({*Pass1to2* → *Pass2to1* → *Goal1*} and {*Pass2to1* → *Pass1to2* → *Pass2to1*}) which are already defined as distribution of subsequences of soccer actions. Figure 4 shows these distributions in the offline Analysis (Fig. 4 - top). In the offline analysis

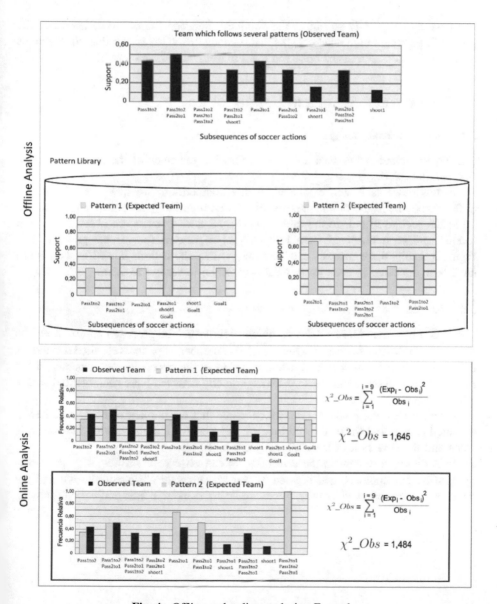

Fig. 4. Offline and online analysis - Example.

(Fig. 4 - down), to decide what play pattern is *more similar* to the behavior of the observed team, two comparisons are done. In this case, we calculate the value of χ^2_Obs for each pattern. The comparison with *Pattern 1* would as follows:

$$\chi^2_{Obs} = \frac{(0,33-0,44)^2}{0,44} + \frac{(0,5-0,5)^2}{0,5} + \frac{(0-0,33)^2}{0,33} + \frac{(0-0,33)^2}{0,33} + \frac{(0,33-0,44)^2}{0,44}$$
$$+ \frac{(0-0,33)^2}{0,33} + \frac{(0-0,16)^2}{0,16} + \frac{(0-0,33)^2}{0,33} + \frac{(0-0,11)^2}{0,1} = 1,645$$

If we apply χ^2_Obs to obtain the similarity between the observed team and the pattern 2, the value obtained is: 1,484. Thus, since this value is less than the previous one, it is considered that the observed team is following the pattern 2.

5   Experimental Setup and Results

5.1   Experimental Design

The experimentation has been doing by following the rules of the *RoboCup Coach Competition 2006*.

In this case, as it was evaluated in that competition, we have considered two different rounds. In the offline analysis of the first round 17 patterns are analyzed, and 16 in the second round. In the online analysis, 3 soccer teams are observed in both rounds, and each observed team follows 4 or 5 patterns (out of the previous 17). Figure 5 details these patterns and how many times the actions related to the pattern have been executed in each team, what is essential to recognize them.

5.2   Results

Figure 6 shows the patterns detected by our approach. The proposed approach results those patterns with a lowest χ^2_Obs value. The number of patterns resulted depend on the number of patterns that have been activated in the observed soccer game. In the Fig. 6, we can observe that those patterns correctly detected have been marked with a red rectangle.

By observing these results, we can conclude that the number of times a pattern is executed is essential for its recognition. We can observe that the relation between this value and the detection of the pattern is very high.

In addition, to complete these results, we can observe the different patterns recognized by our approach and ordered by the χ^2_Obs value. We can observe that the patterns that are followed by the observed team are marked in bold and italic (Table 1).

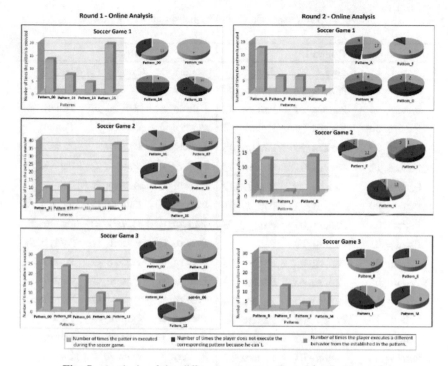

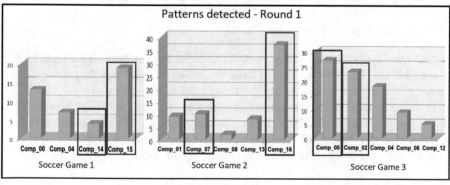

Fig. 5. Analysis of the different patterns activated in the 2 rounds.

Online Analysis: Pattern Detection - Results

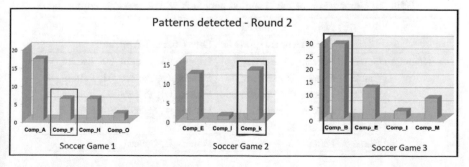

Fig. 6. Results – Patterns detected in each soccer game of the two rounds.

Table 1. Table captions should be placed above the tables.

Round 1			Round 2		
Game 1	Game 2	Game 3	Game 1	Game 2	Game 3
Pattern_04	*Pattern_16*	*Pattern_04*	*Pattern_J*	Pattern_B	*Pattern_F*
Pattern_16	*Pattern_01*	*Pattern_02*	Pattern_K	Pattern_F	Pattern_E
Pattern_00	Pattern_00	Pattern_13	*Pattern_F*	Pattern_A	Pattern_J
Pattern_15	*Pattern_13*	Pattern_05	Pattern_E	*Pattern_E*	*Pattern_B*
Pattern_12	Pattern_05	*Pattern_00*	*Pattern_A*	Pattern_J	Pattern_D
Pattern_05	*Pattern_07*	*Pattern_12*	Pattern_D	*Pattern_K*	Pattern_H
Pattern_09	Pattern_03	Pattern_01	Pattern_G	Pattern_D	Pattern_A
Pattern_03	Pattern_09	*Pattern_06*	Pattern_P	*Pattern_P*	Pattern_P
Pattern_01	Pattern_10	Pattern_03	Pattern_C	Pattern_H	*Pattern_M*
Pattern_06	Pattern_06	Pattern_07	Pattern_N	Pattern_C	Pattern_C
Pattern_08	*Pattern_08*	Pattern_10	Pattern_M	Pattern_O	Pattern_K
Pattern_13	Pattern_15	Pattern_16	Pattern_K	Pattern_G	Pattern_O
Pattern_10	Pattern_02	Pattern_11	Pattern_B	Pattern_M	Pattern_G
Pattern_11	Pattern_12	Pattern_08	*Pattern_H*	Pattern_N	Pattern_N
Pattern_14	Pattern_04	Pattern_15	Pattern_I	*Pattern_I*	*Pattern_I*
Pattern_02	Pattern_11	Pattern_09	*Pattern_O*	Pattern_L	Pattern_L
Pattern_07	Pattern_14	Pattern_14			

5.3   Comparative

As we have already detailed, the authors of this paper already proposed an approach for the competition [1]. However, in that case, the online analysis was very different since it was based on the comparison of *tries* using a novel algorithm.

In Table 2, we compare the approach proposed in this paper (*Our Approach*) and the approach proposed in [1] (*Our Previous Approach*). We can observe that our approach performs better than the previous one. In addition, the proposed approach is faster than the previous one and it is not necessary to define any threshold value.

Table 2. Comparison of the proposed approach with a previous approach.

		Patterns recognized	
		Our Previous Approach	**Our Current Approach**
Round 1	Game 1	Pattern_15, Pattern_14	Pattern_04, Pattern_00, Pattern_15
	Game 2	Pattern_16, Pattern_07	Pattern_16, Pattern_01, Pattern_13
	Game 3	Pattern_00, Pattern_02	Pattern_04, Pattern_02, Pattern_00
Round 2	Game 1	Pattern_F, Pattern_J	Pattern_A, Pattern_F, Pattern_J
	Game 2	Pattern_K	Pattern_E
	Game 3	Pattern_B	Pattern_B, Pattern_E

6 Conclusions and Future Work

The research work proposed in this paper is related with a RoboCup Coach Competition which was held more than a decade ago, in 2006. However, the problem proposed in that competition (opponent modeling) is a hot topic in very different environments as it can be seen in the related work. This is due to the importance of adaptation and learning abilities for an agent that intercepts with other selfish agents. In this sense, a soccer game is a specific but very attractive real time multi-agent environment from the point of view of artificial intelligence and multi-agent research.

In this paper, we have presented an approach to tackle the goal proposed in the RoboCup Soccer Simulated. The approach is based on the treatment of a team behavior as a sequence of high-level soccer actions. Thus, after obtaining the corresponding sequences of actions that define a soccer team, these are treated and analyzed to obtain the most relevant subsequences that define a play pattern. Finally, we propose a statistical method for detecting the patterns followed by an observed team.

In this research, we have considered that the behavior of a team does not change over a game, but the construction of soccer team models could consider the dynamism of the teams to keep the models up to date.

Acknowledgments. This work has been supported by the Spanish Government under projects TRA2015-63708-R and TRA2016-78886-C3-1-R.

References

1. Kitano, H., Tambe, M., Stone, P., Veloso, M., Coradeschi, S., Osawa, E., Matsubara, E., Noda, I., Asada, M.: The RoboCup synthetic agent challenge 97. In: RoboCup-97: Robot Soccer World Cup I (1998)
2. RoboCup Federation: RoboCup Web Page (2018). http://www.robocup.org/
3. Prokopenko, M., Wang, P., Marian, S., Bai, A., Li, X., Chen, X.: RoboCup 2D Soccer Simulation League: Evaluation Challenges. CoRR (2017)
4. Riley, P., Veloso, M.: Coaching a simulated soccer team by opponent model recognition. In: Proceedings of the Fifth International Conference on Autonomous Agents (2001)
5. Noda, I., Matsubara, H., Hiraki, K.: Soccer server: a tool for research. In: Applied AI, pp. 233–258 (1998)
6. Iglesias, J.A., Ledezma, A., Sanchis, A.: The RoboCup Agent Behavior Modeling Challenge. In: Proceedings of the XI Workshop of Physical Agents 2010 (WAF-2010), pp. 179–186 (2010)
7. Reis, L.P., Almeida, F., Mota, L., Lau, N.: Coordination in multi-robot systems: applications in robotic soccer. In: International Conference on Agents and Artificial Intelligence (2012)
8. Almeida, F., Abreu, P.H., Lau, N., Reis, L.P.: An automatic approach to extract goal plans from soccer simulated matches. Soft. Comput. **17**(5), 835–848 (2013)
9. Abreu, P.H., Silva, D.C., Almeida, F., Mendes-Moreira, J.: Improving a simulated soccer team's performance through a Memory-Based Collaborative Filtering approach. Appl. Soft Comput. **23**, 180–193 (2014)
10. Tavcar, A., Kuznar, D., Gams, M.: Hybrid Multi-Agent Strategy Discovering Algorithm for human behavior. Expert Syst. Appl. **71**, 370–382 (2017)

11. Carmel, D., Markovitch, S.: Incorporating opponent models into adversary search. In: AAAI/IAAI, vol. 1 (1996)
12. Carmel, D., Markovitch, S.: Opponent modeling in multi-agent systems. In: International Joint Conference on Artificial Intelligence (1995)
13. Larik, A.S., Haider, S.: Opponent classification in robot soccer. In: International Conference on Industrial, Engineering and Other Applications of Applied Intelligent Systems (2015)
14. He, H., Boyd-Graber, J., Kwok, K., Daume III, H.: Opponent modeling in deep reinforcement learning. In: International Conference on Machine Learning (2016)
15. Iglesias, J.A., Ledezma, A., Sanchis, A.: Caos coach 2006 simulation team: an opponent modelling approach. Comput. Inform. **28**(1), 57–80 (2012)
16. Iglesias, J.A., Palacin, M.A., Ledezma, A., Sanchis, A.: VIENA: VIsual ENvironment for Analyzing Robocup Soccer Teams. In: Proceedings of the X Simposio de Inteligencia Computacional (SICO 2010) (2010)
17. Iglesias, J.A., Fernandez, J.A., Villena, I.R., Ledezma, A., Sanchis, A.: The winning advantage: using opponent models in robot soccer. In: International Conference on Intelligent Data Engineering and Automated Learning (2009)
18. Iglesias, J.A., Ledezma, A., Sanchis, A.: Creating user profiles from a command-line interface: a statistical approach. In: International Conference on User Modeling, Adaptation, and Personalization (2009)
19. Kuhlmann, G., Stone, P., Lallinger, J.: The champion UT Austin Villa 2003 simulator online coach team. In: RoboCup-2003: Robot Soccer World Cup VII, Berlin (2004)
20. Iglesias, J.A., Ledezma, A., Sanchis, A.: Sequence classification using statistical pattern recognition. In: International Symposium on Intelligent Data Analysis (2007)
21. Fredkin, E.: Trie memory. Commun. ACM **3**(9), 490–499 (1960)
22. Agrawal, R., Srikant, R.: Mining sequential patterns. In: Proceedings of the Eleventh International Conference on Data Engineering (1995)

Other Topics

First Steps Towards a General Algorithm to Estimate the Mechanical State of a Vehicle

Pablo Bernal-Polo$^{(\boxtimes)}$ and Humberto Martínez-Barberá

Facultad de Informática, University of Murcia, 30100 Murcia, Spain
{pablo.bernal,humberto}@um.es

Abstract. An orientation estimation algorithm is presented. Multiple Inertial Measurement Units (IMUs) are used as sources of information. Both the state of the system, and the position and orientation of the sensors within the system are estimated. Quaternions are adopted as the orientation representation. The algorithm is developed under the Extended Kalman Filter formalism. An experiment is used to evaluate the sensor fusion algorithm. This work aims to be the preamble of a more general estimation algorithm that merges the information from multiple sensors.

Keywords: Sensor fusion · Kalman filter · Quaternions · IMUs

1 Introduction

Sensor fusion is a research area of growing interest. In particular, sensor fusion for estimating the mechanical state of a system is attractive due to the importance of its applications. It is valuable in navigation because there are situations in which a GPS could not give adequate measurements. For example, when the signal of the satellite is lost due to blocking or attenuation, or when the measurements are corrupted due to bouncing with nearby structures. In such cases, the addition of other sensors is beneficial for the estimation of the position. In the case of autonomous vehicles, an accurate estimation of the mechanical state is mandatory. To design their control algorithms it is necessary to know precisely their orientation, angular velocity, position, velocity, and sometimes some extra information about their environment. Some other applications in which mechanical state estimation is imperative are virtual reality, augmented reality, or tracking.

Although there are other tools used for estimating the mechanical state of a system, the Kalman filter [2] provides a convenient framework for sensor fusion. Unit quaternions can be used as the orientation representation, and the Kalman filter can be adapted to produce estimations based on them. Knowing the advantages of unit quaternions over other orientation representations (low dimensionality, continuous, non-singular, linear motion equations), this is a valuable feature.

© Springer Nature Switzerland AG 2019
R. Fuentetaja Pizán et al. (Eds.): WAF 2018, AISC 855, pp. 319–332, 2019.
https://doi.org/10.1007/978-3-319-99885-5_22

This paradigm has led to numerous successful stories regarding sensor fusion. It was first approached in [3] were they fused information from gyroscopes and attitude sensors. After them, others have tackled the problem, acquiring a deep understanding of the technique [1,5]. The increase in computational speeds allows large amounts of information to be processed. For example, Inertial Measurement Units (IMUs) provide acceleration and angular velocity measurements at rates up to 1000 Hz. It is also possible to include more complex sensors such as cameras that produce information-rich measurements. [4,6–8] are examples of IMU-camera fusion in which the state of the system is estimated, but also the relative orientation and position between sensors. It is necessary to know this information to establish a common reference frame between these sensors, but in doing so, the dimension of the state is also enlarged. That entails an increase in the computational cost, despite information such as the position and orientation of a sensor may not be correlated with the dynamic state of the system. In addition, all of them use the IMU measurements as control inputs for the prediction of the state, which implies not being able to merge the information provided by multiple IMUs in a direct way.

In this paper we explore the idea of merging the information of multiple sensors mounted on a vehicle. We assume that each sensor has a fixed position and orientation within the system. We also separate the state of the system and the state of each sensor in an independent random variable. Thus, each time a measurement arrives, we only update the state of the system and the state of the sensor that provided the measurement. In this way, the number of sensors would only be restricted by the speed of our processor, and not by the quadratic growth of the covariance matrix due to the increase in the size of a single state of the system. We use four IMUs as sensors to test the concept, and we consider that their measurements are measurements in the Kalman filter formalism; no control inputs.

The rest of the paper is structured as follows. In Sect. 2 we present the estimation algorithm. In Sect. 3 we show the results of our validation experiment. Section 4 concludes the paper, and proposes new research objectives.

2   Estimation Algorithm

In this section we present the algorithm used to estimate both the state of the system, and the state of the sensors within the system.

2.1   Notation and Formulation

In this subsection we will define the notation used through this paper.

A vector $\mathbf{v}$ expressed in a reference frame $\mathcal{R}$ will be denoted as $^R\mathbf{v}$. If the vector $\mathbf{v}$ describes a property of a reference frame $\mathcal{F}$, we will write $\mathbf{v}^F$. Then, a vector $\mathbf{v}$ that describes a property of a reference frame $\mathcal{F}$, and is expressed in a reference frame $\mathcal{R}$ will be denoted as $^R\mathbf{v}^F$. For example, in Fig. 1, $\mathbf{x}^B$ represents the position of reference frame $\mathcal{B}$ measured from reference frame $\mathcal{O}$. Consequently, if we want the position of reference frame $\mathcal{B}$ measured from reference frame $\mathcal{O}$ expressed in reference frame $\mathcal{O}$, then we will write $^O\mathbf{x}^B$.

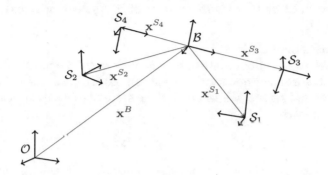

Fig. 1. Representation of several vectors to illustrate the notation used

The expressions of a vector measured in different reference frames are related through a rotation transformation. We will describe rotation transformations using unit quaternions. Each rotation transformation is mapped with a rotation matrix $\mathbf{R}$, and with two unit quaternions $\mathbf{q}$ and $-\mathbf{q}$ through the relation

$$\mathbf{R}(\mathbf{q}) = \begin{pmatrix} 1 - 2q_2^2 - 2q_3^2 & 2(q_1\,q_2 - q_3\,q_0) & 2(q_1\,q_3 + q_2\,q_0) \\ 2(q_1\,q_2 + q_3\,q_0) & 1 - 2q_1^2 - 2q_3^2 & 2(q_2\,q_3 - q_1\,q_0) \\ 2(q_1\,q_3 - q_2\,q_0) & 2(q_2\,q_3 + q_1\,q_0) & 1 - 2q_1^2 - 2q_2^2 \end{pmatrix}. \tag{1}$$

We will denote $^R\mathbf{q}^F$ to the unit quaternion describing the rotation transformation that maps a vector $^F\mathbf{v}$ expressed in reference frame $\mathcal{F}$ with the same vector $^R\mathbf{v}$ expressed in reference frame $\mathcal{R}$. Then, we will have

$$\mathbf{R}(^R\mathbf{q}^F)\ ^F\mathbf{v}\ =\ ^R\mathbf{v}. \tag{2}$$

The orientation of a reference frame is described as a rotation transformation from another reference frame. For example, in Fig. 1, we could describe the orientation of $\mathcal{B}$ as the rotation transformation that maps a vector $^B\mathbf{v}$ expressed in reference frame $\mathcal{B}$, with the same vector $^O\mathbf{v}$ expressed in reference frame $\mathcal{O}$. That rotation transformation would be represented by the unit quaternion $^O\mathbf{q}^B$.

System State Formulation

We will consider an inertial reference frame $\mathcal{O}$, as the one in Fig. 1, with respect to which we will define the mechanical state of our system. Our system will be considered a rigid body with a non-inertial reference frame $\mathcal{B}$ attached to it. We will define the mechanical state of our system through the vector of random variables

$$\mathbf{s}^B\ =\ (\ ^O\mathbf{q}^B\ ,\ ^B\boldsymbol{\omega}^B\ ,\ ^O\mathbf{x}^B\ ,\ ^O\mathbf{v}^B\ ), \tag{3}$$

where $^Oq^B$ represents the unit quaternion that describes the rotation transformation that maps a vector ^Bv measured in reference frame $\mathcal{B}$ to the same vector ^Ov measured in the inertial reference frame $\mathcal{O}$, $^B\omega^B$ is the angular velocity of the system expressed in $\mathcal{B}$, $^Ox^B$ is the position of the system measured from $\mathcal{O}$ and expressed in $\mathcal{O}$, and $^Ov^B$ is the velocity of the system measured from $\mathcal{O}$ and expressed in $\mathcal{O}$.

Sensor State Formulation

To estimate the state of our system, several sensors will be mounted on it. Each sensor will measure some quantity, and will express it in a certain reference frame $\mathcal{S}$ attached to said sensor. We will define the state of each sensor using a vector of random variables

$$\mathbf{s}^S \;\;=\;\; (\; ^Bq^S, \; ^Bx^S \;), \tag{4}$$

where $^Bq^S$ represents the unit quaternion that describes the rotation transformation that maps a vector ^Sv expressed in the sensor reference frame $\mathcal{S}$ to the same vector ^Bv expressed in the system reference frame $\mathcal{B}$.

We will consider the random variables $\mathbf{s}^B$ and $\mathbf{s}^S$ to be independent.

2.2   Model Equations

This subsection contains the equations used to model both the evolution of the system, and the measurements of the sensors.

System Motion Equations

The evolution of the state of the system will be modeled using the following equations:

$$\frac{d\ ^Oq^B(t)}{dt} \;\;=\;\; \frac{1}{2}\ ^Oq^B(t) * {}^B\omega^B(t), \tag{5}$$

$$\frac{d\ ^B\omega^B(t)}{dt} \;\;=\;\; \widetilde{\tau}(t), \tag{6}$$

$$\frac{d\ ^Ox^B(t)}{dt} \;\;=\;\; {}^Ov^B(t), \tag{7}$$

$$\frac{d\ ^Ov^B(t)}{dt} \;\;=\;\; {}^Oa^B(t), \tag{8}$$

where $\widetilde{\tau}$ represents the process noise in angular velocity, and is associated with the torque acting on the system expressed in reference frame $\mathcal{B}$, and its inertia tensor. On the other hand, $^Oa^B$ is the acceleration of the system measured from reference frame $\mathcal{O}$ and expressed in reference frame $\mathcal{O}$, and represents the process noise in the velocity.

Sensor Motion Equations

The evolution of the state of each sensor will be modeled using the equations

$$\frac{d\ ^B q^S(t)}{dt} \quad = \quad \frac{1}{2}\ ^B q^S * q^q, \tag{9}$$

$$\frac{d\ ^B \mathbf{x}^S(t)}{dt} \quad = \quad \mathbf{q}^x, \tag{10}$$

where q^q represents a small process noise in the orientation of the sensor within the system, and $\mathbf{q}^x$ represents a small process noise in the position of the sensor within the system. We will assume that both the expected value of q^q and $\mathbf{q}^x$ are zero.

Measurement Equations

For this work we will only consider one type of sensors: IMUs. We will start deriving the measurement equations that relate the gyroscope measurement and the accelerometer measurement with the state of the system.

Gyroscope Measurement Equation
The gyroscope measurement $\boldsymbol{\omega}^m = {}^S\boldsymbol{\omega}^S$ is related with the state of the system through

$$\mathbf{R}(^B q^S)\ ^S\boldsymbol{\omega}^S \quad = \quad {}^B\boldsymbol{\omega}^B. \tag{11}$$

Then, the gyroscope measurement equation is

$$\boldsymbol{\omega}^m \quad = \quad \mathbf{R}^T(^B q^S)\ ^B\boldsymbol{\omega}^B. \tag{12}$$

Accelerometer Measurement Equation
The accelerometer measurement $\mathbf{a}^m = {}^S\mathbf{a}^S$ is related with the state of the system through

$$\mathbf{R}(^O q^B)\ \mathbf{R}(^B q^S)\ ^S\mathbf{a}^S \quad = \quad {}^O\mathbf{a}^S - {}^O\mathbf{g}^S, \tag{13}$$

where $^O\mathbf{g}^S$ is the acceleration due to gravity expressed in reference frame $\mathcal{O}$, and at the point where the sensor is located. Noticing that $^O\mathbf{x}^S = {}^O\mathbf{x}^B + \mathbf{R}(^O q^B)\ ^B\mathbf{x}^S$ we can compute $^O\mathbf{a}^S = \frac{d^2\ ^O\mathbf{x}^S}{dt^2}$:

$$\frac{d\ ^O\mathbf{x}^S}{dt} = \frac{d\ ^O\mathbf{x}^B}{dt} + \mathbf{R}(^O q^B) \left(\underbrace{\frac{d\ ^B\mathbf{x}^S}{dt}}_{0} + {}^B\boldsymbol{\omega}^B \times {}^B\mathbf{x}^S \right) \quad \Longrightarrow \tag{14}$$

$$\Longrightarrow \quad \frac{d^2\ ^O\mathbf{x}^S}{dt^2} = \frac{d^2\ ^O\mathbf{x}^B}{dt^2} + \mathbf{R}(^O q^B) \left(\frac{d\ ^B\boldsymbol{\omega}^B}{dt} \times {}^B\mathbf{x}^S + \right.$$

$$\left. + {}^B\boldsymbol{\omega}^B \times \underbrace{\frac{d\ ^B\mathbf{x}^S}{dt}}_{0} + {}^B\boldsymbol{\omega}^B \times ({}^B\boldsymbol{\omega}^B \times {}^B\mathbf{x}^S) \right) \quad = \tag{15}$$

$$\implies \quad {}^O\mathbf{a}^S = {}^O\mathbf{a}^B + \mathbf{R}({}^O\mathbf{q}^B)\left(\frac{d\,{}^B\boldsymbol{\omega}^B}{dt} \times {}^B\mathbf{x}^S + \right.$$
$$\left. + {}^B\boldsymbol{\omega}^B \times \left({}^B\boldsymbol{\omega}^B \times {}^B\mathbf{x}^S\right)\right). \tag{16}$$

Then, the accelerometer measurement equation is

$$\mathbf{a}^m \;=\; \mathbf{R}^T({}^B\mathbf{q}^S)\left[\mathbf{R}^T({}^O\mathbf{q}^B)\left({}^O\mathbf{a}^B - {}^O\mathbf{g}^S\right) + \right.$$
$$\left. + \frac{d\,{}^B\boldsymbol{\omega}^B}{dt} \times {}^B\mathbf{x}^S + {}^B\boldsymbol{\omega}^B \times \left({}^B\boldsymbol{\omega}^B \times {}^B\mathbf{x}^S\right)\right]. \tag{17}$$

2.3   State Prediction

Taking the expected value of Eqs. (5) to (8), assuming the random variables ${}^O\mathbf{q}^B$ and $\mathbf{q}^\omega(t) = \int_0^t \tilde{\boldsymbol{\tau}}(\tau)d\tau$ to be independent, and integrating with respect to time, we obtain the evolution equations:

$$^O\overline{\mathbf{q}}^B(t) \;=\; {}^O\overline{\mathbf{q}}^B(0) * \begin{pmatrix} \cos\dfrac{\lVert {}^B\overline{\boldsymbol{\omega}}^B\rVert\, t}{2} \\ \dfrac{{}^B\overline{\boldsymbol{\omega}}^B}{\lVert {}^B\overline{\boldsymbol{\omega}}^B\rVert}\,\sin\dfrac{\lVert {}^B\overline{\boldsymbol{\omega}}^B\rVert\, t}{2} \end{pmatrix}, \tag{18}$$

$$^B\overline{\boldsymbol{\omega}}^B(t) \;=\; {}^B\overline{\boldsymbol{\omega}}^B(0) + \overline{\mathbf{q}}^\omega(t), \tag{19}$$

$$^O\overline{\mathbf{x}}^B(t) \;=\; {}^O\overline{\mathbf{x}}^B(0) + {}^O\overline{\mathbf{v}}^B(0)\,t + {}^O\overline{\mathbf{a}}^B(0)\,\frac{t^2}{2}, \tag{20}$$

$$^O\overline{\mathbf{v}}^B(t) \;=\; {}^O\overline{\mathbf{v}}^B(0) + {}^O\overline{\mathbf{a}}^B(0)\,t. \tag{21}$$

On the other hand, taking the expected value of Eqs. (9) and (10), assuming the random variables ${}^B\mathbf{q}^S$ and $\mathbf{q}^q$ to be independent, and integrating with respect to time, results in the following equations:

$$^B\overline{\mathbf{q}}^S(t) \;=\; {}^B\overline{\mathbf{q}}^S(0), \tag{22}$$

$$^B\overline{\mathbf{x}}^S(t) \;=\; {}^B\overline{\mathbf{x}}^S(0). \tag{23}$$

We have found the evolution equations for the expected values. Now we need to find the evolution equations for the covariance matrix. For that purpose, we need to find how the random variables

$$\Delta\mathbf{s}^B \;=\; \left({}^O\mathbf{e}^B,\ \Delta{}^B\boldsymbol{\omega}^B,\ \Delta{}^O\mathbf{x}^B,\ \Delta{}^O\mathbf{v}^B\right), \tag{24}$$

$$\Delta\mathbf{s}^S \;=\; \left({}^B\mathbf{e}^S,\ \Delta{}^B\mathbf{x}^S\right), \tag{25}$$

evolve, where for a vector $\mathbf{v}$ we have denoted $\Delta\mathbf{v} = \mathbf{v} - \overline{\mathbf{v}}$, and $^{O}\mathbf{e}^{B}$ and $^{B}\mathbf{e}^{S}$ are the random variables that represent the orientation in the charts centered in $^{O}\overline{q}^{B}$ and $^{B}\overline{q}^{S}$ respectively, as was done in [1]. Using differential Eqs. (5) to (8), and evolution Eqs. (18) to (21), we obtain

$$\frac{d\,\Delta\mathbf{s}^{B}}{dt} \approx \begin{pmatrix} \mathbf{I} & -\left[^{B}\overline{\boldsymbol{\omega}}^{B}\right]_{\times} & \mathbf{0} & \mathbf{0} \\ \mathbf{0} & \mathbf{0} & \mathbf{0} & \mathbf{0} \\ \mathbf{0} & \mathbf{0} & \mathbf{I} & \mathbf{0} \\ \mathbf{0} & \mathbf{0} & \mathbf{0} & \mathbf{0} \end{pmatrix} \Delta\mathbf{s}^{B} + \begin{pmatrix} \mathbf{0} \\ \Delta\mathbf{q}^{\omega} \\ \mathbf{0} \\ \Delta^{O}\mathbf{a}^{B} \end{pmatrix}, \qquad (26)$$

where we have written $\left[\mathbf{v}\right]_{\times}$ to denote

$$\left[\mathbf{v}\right]_{\times} = \begin{pmatrix} 0 & -v_{3} & v_{2} \\ v_{3} & 0 & -v_{1} \\ -v_{2} & v_{1} & 0 \end{pmatrix}. \qquad (27)$$

On the other hand, using differential Eqs. (9) and (10), and evolution Eqs. (22) and (23), we get

$$\frac{d\,\Delta\mathbf{s}^{S}}{dt} = \begin{pmatrix} \Delta\mathbf{q}^{q} \\ \Delta\mathbf{q}^{x} \end{pmatrix}. \qquad (28)$$

Now, we know that the covariance matrix satisfies the following differential equation (see [9]):

$$\frac{d\,\mathbf{P}}{dt} = \mathbf{F}\mathbf{P} + \mathbf{P}\mathbf{F}^{T} + \mathbf{G}\mathbf{Q}\mathbf{G}^{T}, \qquad (29)$$

where $\mathbf{F}$ and $\mathbf{G}$ are the matrices that model the dynamics of the random variable $\Delta\mathbf{s}$ such that $\mathbf{P} = \mathrm{E}\left[(\Delta\mathbf{s})^{T}\Delta\mathbf{s}\right]$:

$$\frac{d\,\Delta\mathbf{s}}{dt} = \mathbf{F}\,\Delta\mathbf{s} + \mathbf{G}\,\mathbf{q}. \qquad (30)$$

Comparing Eq. (30) with (26) and (28) we identify the matrices $\mathbf{F}$ and $\mathbf{G}$ that appear in the differential equations for $\mathbf{P}^{B} = \mathrm{E}\left[(\Delta\mathbf{s}^{B})^{T}\Delta\mathbf{s}^{B}\right]$ and $\mathbf{P}^{S} = \mathrm{E}\left[(\Delta\mathbf{s}^{S})^{T}\Delta\mathbf{s}^{S}\right]$. The resulting differential equations can be approximately solved:

$$\mathbf{P}^{B}(t) \approx e^{\mathbf{F}t}\left(\mathbf{P}^{B}(0) + \mathbf{Q}^{B}\,t\right)e^{\mathbf{F}^{T}t}, \qquad (31)$$
$$\mathbf{P}^{S}(t) \approx \mathbf{Q}^{S}\,t, \qquad (32)$$

with

$$e^{\mathbf{F}t} = \begin{pmatrix} \mathbf{R}^{T}\left(^{B}\boldsymbol{\omega}^{B}\,t\right) & \mathbf{I}\,t & \mathbf{0} & \mathbf{0} \\ \mathbf{0} & \mathbf{I} & \mathbf{0} & \mathbf{0} \\ \mathbf{0} & \mathbf{0} & \mathbf{I} & \mathbf{I}\,t \\ \mathbf{0} & \mathbf{0} & \mathbf{0} & \mathbf{I} \end{pmatrix}, \qquad (33)$$

$$\mathbf{Q}^B \;\; = \;\; \begin{pmatrix} 0 & 0 & 0 & 0 \\ 0 & \mathbf{Q}^\omega & 0 & 0 \\ 0 & 0 & 0 & 0 \\ 0 & 0 & 0 & \mathbf{Q}^a \end{pmatrix}, \tag{34}$$

$$\mathbf{Q}^S \;\; = \;\; \begin{pmatrix} \mathbf{Q}^q & 0 \\ 0 & \mathbf{Q}^x \end{pmatrix}, \tag{35}$$

being $\mathbf{Q}^\omega = \mathrm{E}\left[(\Delta\mathbf{q}^\omega)^T \Delta\mathbf{q}^\omega\right]$, $\mathbf{Q}^a = \mathrm{E}\left[(\Delta^O\mathbf{a}^B)^T \Delta^O\mathbf{a}^B\right]$, $\mathbf{Q}^q = \mathrm{E}\left[(\Delta\mathbf{q}^q)^T \Delta\mathbf{q}^q\right]$, and $\mathbf{Q}^x = \mathrm{E}\left[(\Delta\mathbf{q}^x)^T \Delta\mathbf{q}^x\right]$.

2.4   Measurement Prediction

Taking expected values in Eqs. (12) and (17), assuming that $\overline{\overline{\tau}} = \mathbf{0}$, and neglecting moments of order greater than 1 we obtain the expected measurement:

$$\overline{\mathbf{z}} \;\; = \;\; \begin{pmatrix} \overline{\mathbf{a}}^m \\ \overline{\boldsymbol{\omega}}^m \end{pmatrix} \;\; = \;\; \begin{pmatrix} \mathbf{R}^T(^B\overline{\boldsymbol{q}}^S) \left[{}^B\widetilde{\mathbf{a}}^B + {}^B\overline{\boldsymbol{\omega}}^B \times \left({}^B\overline{\boldsymbol{\omega}}^B \times {}^B\overline{\mathbf{x}}^S \right) \right] \\ \mathbf{R}^T(^B\overline{\boldsymbol{q}}^S) \; {}^B\overline{\boldsymbol{\omega}}^B \end{pmatrix}, \tag{36}$$

where ${}^B\widetilde{\mathbf{a}}^B = \mathbf{R}^T(^O\overline{\boldsymbol{q}}^B)\left({}^O\overline{\mathbf{a}}^B - {}^O\overline{\mathbf{g}}^S \right)$.

The covariance matrix of the measurement prediction is computed through

$$\mathbf{S} \;\; = \;\; \mathbf{H}^B \, \mathbf{P}^B \left(\mathbf{H}^B\right)^T + \mathbf{H}^S \, \mathbf{P}^S \left(\mathbf{H}^S\right)^T, \tag{37}$$

where $\mathbf{H}^B = \frac{d\mathbf{z}}{d\mathbf{s}^B}(\overline{\mathbf{s}}^B, \overline{\mathbf{s}}^S)$ and $\mathbf{H}^S = \frac{d\mathbf{z}}{d\mathbf{s}^S}(\overline{\mathbf{s}}^B, \overline{\mathbf{s}}^S)$. In our particular case these matrices take the form

$$\mathbf{H}^B \;\; = \;\; \begin{pmatrix} \mathbf{R}^T(^B\overline{\boldsymbol{q}}^S)\left[{}^B\widetilde{\mathbf{a}}^B\right]_\times & \mathbf{R}^T(^B\overline{\boldsymbol{q}}^S)\frac{d\,[{}^B\boldsymbol{\omega}^B]^2_\times \, {}^B\mathbf{x}^S}{d\,{}^B\boldsymbol{\omega}^B}(\overline{\mathbf{s}}^B,\overline{\mathbf{s}}^S) & 0 & 0 \\ 0 & \mathbf{R}^T(^B\overline{\boldsymbol{q}}^S) & 0 & 0 \end{pmatrix}, \tag{38}$$

$$\mathbf{H}^S \;\; = \;\; \begin{pmatrix} \left[\overline{\mathbf{a}}^m\right]_\times & \mathbf{R}^T(^B\overline{\boldsymbol{q}}^S)\left[{}^B\overline{\boldsymbol{\omega}}^B\right]^2_\times \\ \left[\overline{\boldsymbol{\omega}}^m\right]_\times & 0 \end{pmatrix}, \tag{39}$$

with

$$\frac{d\,[{}^B\boldsymbol{\omega}^B]^2_\times \, {}^B\mathbf{x}^S}{d\,{}^B\boldsymbol{\omega}^B} \;\; = \;\; {}^B\boldsymbol{\omega}^B \, ({}^B\mathbf{x}^S)^T - 2\,{}^B\mathbf{x}^S \, ({}^B\boldsymbol{\omega}^B)^T + \left(({}^B\mathbf{x}^S)^T \, {}^B\boldsymbol{\omega}^B \right) \mathbf{I}. \tag{40}$$

2.5   State Update

Once we have the state covariance matrices $\mathbf{P}^B$ and $\mathbf{P}^S$, the observation matrices $\mathbf{H}^B$ and $\mathbf{H}^S$ and the measurement covariance matrix $\mathbf{S}$, and remembering definition (24), we can update the system state through

$$\mathbf{K}_n^B \;\; = \;\; \mathbf{P}_{n|n-1}^B \left(\mathbf{H}_n^B\right)^T \mathbf{S}_{n|n-1}^{-1}, \tag{41}$$

$$\Delta \mathbf{s}_{n|n}^{B} \quad = \quad \mathbf{K}_{n}^{B} \left(\mathbf{z}_{n} - \bar{\mathbf{z}}_{n|n-1} \right), \tag{42}$$

$$\mathbf{s}_{n|n}^{B} \quad = \quad \begin{pmatrix} {}^{O}\mathbf{q}^{B}{}_{n|n} = {}^{O}\mathbf{q}^{B}{}_{n|n-1} * \boldsymbol{\delta}({}^{O}\mathbf{e}^{B}{}_{n|n}) \\ {}^{B}\boldsymbol{\omega}^{B}{}_{n|n} = {}^{B}\boldsymbol{\omega}^{B}{}_{n|n-1} + \Delta^{B}\boldsymbol{\omega}^{B}{}_{n|n} \\ {}^{O}\mathbf{x}^{B}{}_{n|n} = {}^{O}\mathbf{x}^{B}{}_{n|n-1} + \Delta^{O}\mathbf{x}^{B}{}_{n|n} \\ {}^{O}\mathbf{v}^{B}{}_{n|n} = {}^{O}\mathbf{v}^{B}{}_{n|n-1} + \Delta^{O}\mathbf{v}^{B}{}_{n|n} \end{pmatrix}, \tag{43}$$

$$\mathbf{P}_{n|n}^{B} \quad = \quad \left(\mathbf{I} - \mathbf{K}_{n}^{B} \mathbf{H}_{n}^{B} \right) \mathbf{P}_{n|n-1}^{B}, \tag{44}$$

Similarly, we can compute the sensor state update through

$$\mathbf{K}_{n}^{S} \quad = \quad \mathbf{P}_{n|n-1}^{S} (\mathbf{H}_{n}^{S})^{T} \mathbf{S}_{n|n-1}^{-1}, \tag{45}$$

$$\Delta \mathbf{s}_{n|n}^{S} \quad = \quad \mathbf{K}_{n}^{S} \left(\mathbf{z}_{n} - \bar{\mathbf{z}}_{n|n-1} \right), \tag{46}$$

$$\mathbf{s}_{n|n}^{S} \quad = \quad \begin{pmatrix} {}^{B}\mathbf{q}^{S}{}_{n|n} = {}^{B}\mathbf{q}^{S}{}_{n|n-1} * \boldsymbol{\delta}({}^{B}\mathbf{e}^{S}{}_{n|n}) \\ {}^{B}\mathbf{x}^{S}{}_{n|n} = {}^{B}\mathbf{x}^{S}{}_{n|n-1} + \Delta^{B}\mathbf{x}^{S}{}_{n|n} \end{pmatrix}, \tag{47}$$

$$\mathbf{P}_{n|n}^{S} \quad = \quad \left(\mathbf{I} - \mathbf{K}_{n}^{S} \mathbf{H}_{n}^{S} \right) \mathbf{P}_{n|n-1}^{S}. \tag{48}$$

Note that the updates for the expected values are not the equations that appear in the classical formulation of the Kalman filter. In fact, we have made use of the manifold update exposed in [1].

3   Experimental Validation

In this section, we present the validation experiment and expose its results. The experiment consists in estimating the state of a real multisensory system. As stated in Subsect. 2.1, the state is composed of two parts: the dynamical state of the system, and the positions and orientations of the sensors. The algorithm will start from a total ignorance of the state of the system, and will incorporate the information provided by the sensor measurements to obtain an estimation of both the dynamical state of the system, and the positions and orientations of the sensors. The objective of the experiment is to obtain a first measure of the accuracy of the estimations produced by the algorithm. In particular, we will compare the real and estimated positioning of the sensors within the system. In the following subsections we will first describe the experiment in detail. Next, we will present its results.

Fig. 2. System used in the experiment

3.1 Description of the Experiment

Four IMUs are mounted on the corners of a protoboard as shown in Fig. 2.

After measuring the distance between the IMUs, we can report that they are located in the corners of a rectangle of sides 15 and 5 cm approximately. All z axes point down, and x axes point outward in a direction orthogonal to the longest side of the rectangle; the y axes point in the direction that makes the reference frame anchored to the IMUs right-handed. Each IMU is connected to an Arduino Nano that requests measurements through I2C. When an Arduino Nano receives a measurement, it is sent using the serial port to a Raspberry Pi 3. The Raspberry Pi 3 sends all the measurements it receives via the Internet to a computer. The computer receives the measurements and merges the information into the system and sensor state using the estimation algorithm described in Sect. 2.

We will generate a random trajectory with the system, and we will store the measurements obtained with each sensor. Then, we will use our algorithm to process the stored measurements, and to obtain the estimated state of the system (dynamical state, and positions and orientations of the sensors). Finally, we will compare the real positioning of the IMUs in the protoboard, and the one estimated by the algorithm.

3.2 Results

We processed the measurements obtained during the random trajectory with our estimation algorithm. Figure 3 displays the estimated orientation represented by the unit quaternion $^O\overline{q}^B$, and angular velocity $^B\overline{\omega}^B$, as a function of time. Figure 3a shows how the variance in each component of the quaternion decreases as time passes, which means a greater precision in the estimation of the orientation of the system.

Estimated orientation of the body VS time

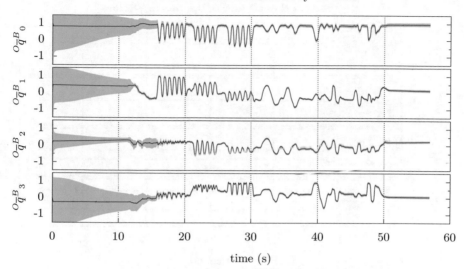

(a) Estimated orientation of the system represented by the unit quaternion $^O\overline{q}^B$

Estimated angular velocity of the body VS time

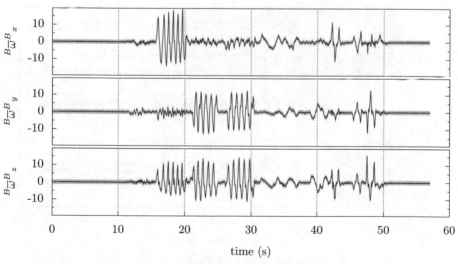

(b) Estimated angular velocity of the system $^B\overline{\omega}^B$ $(\mathrm{rad \cdot s^{-1}})$

Fig. 3. Estimated orientation and angular velocity of the system. The filled region represents the uncertainty in the prediction (5σ)

Estimated orientation of the sensors VS time

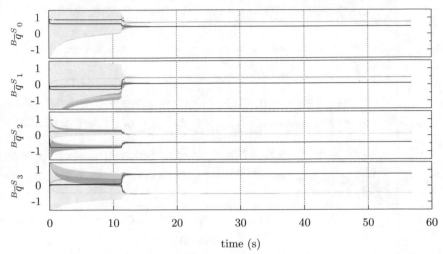

(a) Estimated orientation of the sensors represented by the unit quaternions $\{{}^{B}\overline{\boldsymbol{q}}^{S}{}_{i}\}_{i}$

Estimated position of the sensors VS time

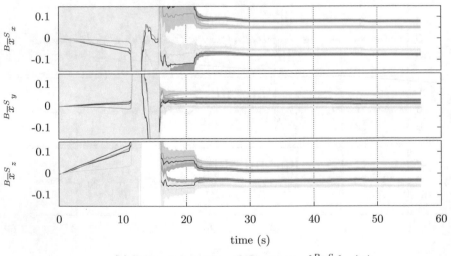

(b) Estimated position of the sensors $\{{}^{B}\overline{\mathbf{x}}^{S}{}_{i}\}_{i}$ (m)

Fig. 4. Estimated orientation and position of the sensors within the system. The filled region represents the uncertainty in the prediction (5σ)

Figure 4 displays the estimated orientation of the sensors $\{^{B}\overline{q}^{S}{}_{i}\}_{i}$, and their position $\{^{B}\overline{\mathbf{x}}^{S}{}_{i}\}_{i}$ within the system as a function of time. As in Fig. 3a, the variance decreases as more measurements are processed. In the end the variance is small enough to be confident with our predictions. However, in Fig. 4 it is difficult to see if the estimation corresponds to the actual arrangement of the sensors. We can transform the estimated positions from $^{B}\overline{\mathbf{x}}^{S}$ to $^{S_1}\overline{\mathbf{x}}^{S}$ through

$$^{S_1}\overline{\mathbf{x}}^{S} \quad = \quad \mathbf{R}^{T}(^{B}\overline{q}^{S_1})\left(^{B}\overline{\mathbf{x}}^{S} - {}^{B}\overline{\mathbf{x}}^{S_1}\right). \tag{49}$$

Figure 5 shows the estimated position of the sensors measured from $\mathcal{S}_1$ and expressed in $\mathcal{S}_1$.

We can appreciate how the estimation produced by our algorithm converges towards the values that were expected taking into account the real disposition of the sensors. Table 1 offers a comparison of the values that were expected with the values estimated by the algorithm.

Estimated position of the sensors VS time

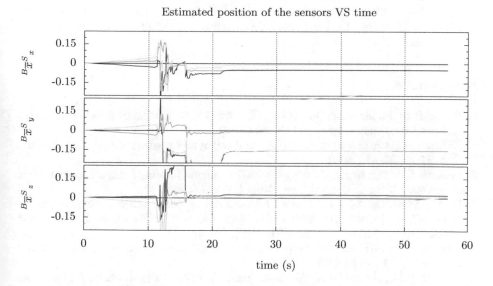

Fig. 5. Estimated position of the sensors measured from $\mathcal{S}_1$ and expressed in $\mathcal{S}_1$ (m)

Table 1. Numerical comparison of the expected position of the sensors measured from $\mathcal{S}_1$ and expressed in $\mathcal{S}_1$, and the one estimated by the algorithm.

Sensor	Expected position (cm)	Estimated position (cm)
S_1	$(0, 0, 0)$	$(0, 0, 0)$
S_2	$(-5, 0, 0)$	$(-5.0, 0.2, 0.2)$
S_3	$(-5, -15, 0)$	$(-4.8, -15.5, 2.4)$
S_4	$(0, -15, 0)$	$(-0.9, -15.6, 1.7)$

4   Conclusion and Future Work

We have explored the idea of merging information from multiple sensors within a vehicle. We have started by trying to merge the information of multiple IMUs using an Extended Kalman filter. The state of the vehicle has been separated from the state of the sensors, and the Kalman filter has been designed to update the state using the measurements of each sensor asynchronously. This approach has several advantages over the classical method in which there is a single state that comprises all the information:

- It allows to update the state of each sensor separately, what relieves the computational cost.
- It establishes a base to be able to include an unlimited number of sensors.

We have tested the algorithm with a validation experiment, and the results look promising.

Having succeed with our approach, the next step would be to include more type of sensors in our algorithm such as GPS, speedometers, or cameras. The quality of the estimations will also be analyzed in detail by performing more experiments.

References

1. Bernal-Polo, P., Martínez Barberá, H.: Orientation estimation by means of extended Kalman filter, quaternions, and charts (2017)
2. Kalman, R.E.: A new approach to linear filtering and prediction problems. J. Basic Eng. **82**(1), 35–45 (1960)
3. Lefferts, E.J., Markley, F.L., Shuster, M.D.: Kalman filtering for spacecraft attitude estimation. J. Guid. Control Dyn. **5**(5), 417–429 (1982)
4. Li, M., Mourikis, A.I.: 3-D motion estimation and online temporal calibration for camera-IMU systems. In: 2013 IEEE International Conference on Robotics and Automation (ICRA), pp. 5709–5716. IEEE (2013)
5. Markley, F.L.: Attitude error representations for Kalman filtering. J. Guid. Control Dyn. **26**(2), 311–317 (2003)
6. Mirzaei, F.M., Roumeliotis, S.I.: A Kalman filter-based algorithm for IMU-camera calibration: observability analysis and performance evaluation. IEEE Trans. Robot. **24**(5), 1143–1156 (2008)
7. Mourikis, A.I., Roumeliotis, S.I.: A multi-state constraint Kalman filter for vision-aided inertial navigation. In: 2007 IEEE International Conference on Robotics and Automation, pp. 3565–3572. IEEE (2007)
8. Shkurti, F., Rekleitis, I., Scaccia, M., Dudek, G.: State estimation of an underwater robot using visual and inertial information. In: 2011 IEEE/RSJ International Conference on Intelligent Robots and Systems (IROS), pp. 5054–5060. IEEE (2011)
9. Xie, L., Popa, D., Lewis, F.L.: Optimal and Robust Estimation: With an Introduction to Stochastic Control Theory. CRC Press, Boca Raton (2007)

Author Index

© Springer Nature Switzerland AG 2019
R. Fuentetaja Pizán et al. (Eds.): WAF 2018, AISC 855, pp. 333–334, 2019.
https://doi.org/10.1007/978-3-319-99885-5

Printed in the United States
By Bookmasters